Principles of Lasers: Engineering Fundamentals

Principles of Lasers: Engineering Fundamentals

Kayden Robertson

Willford Press
www.willfordpress.com

Published by Willford Press,
118-35 Queens Blvd., Suite 400,
Forest Hills, NY 11375, USA

ISBN: 978-1-68285-739-7

Cataloging-in-Publication Data

Principles of lasers : engineering fundamentals / Kayden Robertson.
p. cm.
Includes bibliographical references and index.
ISBN 978-1-68285-739-7
1. Lasers in engineering. 2. Lasers. 3. Engineering. I. Robertson, Kayden.
TA367.5 .P75 2019
621.366--dc23

For information on all Willford Press publications
visit our website at www.willfordpress.com

WILLFORD PRESS

Printed in China.

Contents

Preface

Laser is a device that amplifies the light produced due to stimulated emission of radiation and then emits light that is spatially and temporally coherent. Lasers operate in either pulsed mode or continuous mode. They are of various types, such as gas lasers, solid-state lasers, semiconductor lasers, dye lasers, fiber lasers, etc. These are used in laser printers, barcode scanners, optical disk drives, laser surgery, DNA sequencing, free-space optical communication, etc. This book outlines the processes and applications of lasers in detail. Such selected concepts that redefine this field have been presented herein. This book is meant for students who are looking for an elaborate reference on lasers.

A foreword of all chapters of the book is provided below:

Chapter 1- A laser is a device, which emits amplified light produced as a result of stimulated emission of electromagnetic radiation. The study of lasers is under the domain of laser science. This chapter explores the fundamentals of lasers and laser science;

Chapter 2- Chemical laser is a laser that is powered by a chemical reaction. These are useful in industrial operations of cutting and drilling. This chapter has been carefully written to provide an easy understanding of the varied aspects of chemical lasers. It includes topics like chemical oxygen iodine laser, all gas-phase iodine laser, HF (DF) laser, MIRACL, etc.;

Chapter 3- A solid-state laser is any laser that uses a solid gain medium. The diverse types of solid state lasers such as diode-pumped solid-state laser, thin-disk laser, Er:YAG laser, ruby laser, fiber laser, dye laser, etc. have been thoroughly discussed in this chapter;

Chapter 4- A gas laser is a laser, which emits coherent light produced by the discharge of electric current through a gas. Many different gases can be used in gas lasers. This chapter closely examines the varied types of gas lasers such as helium–neon laser, carbon dioxide laser, argon ion laser, krypton ion laser, etc. for an extensive understanding of the subject;

Chapter 5- A laser diode is a semiconductor laser. It is a semiconductor device in which the laser beam is created at the junction of the diode. Semiconductor lasers are compact-sized and cheap lasers. This chapter discusses in detail semiconductor lasers and laser diodes and includes topics such as quantum dot laser, quantum cascade laser, laser diode, vertical-cavity surface-emitting laser, etc.;

Chapter 6- There has been significant development in the field of laser technology. The aim of this chapter is to explore the diverse laser devices, such as laser tracker, 3D scanner, laser projector and pointer and their applications in printing, scanning, use in medicine, etc.;

Chapter 7- Machining is a set of processes in which a raw material is cut into a specific shape and size through a controlled material-removal process. The various applications of lasers in machining have been explored in this chapter, such as laser beam machining, laser beam welding, hybrid laser welding, laser shock peening, laser engraving, etc.;

Chapter 8 - Laser cutting is the technology of cutting materials using lasers for industrial manufacturing. The topics elaborated in this chapter which include laser cutting machine, laser vaporization cutting, flame cutting, laser fusion cutting, laser oxygen cutting, etc. help to provide an extensive understanding of laser cutting technology.

At the end, I would like to thank all the people associated with this book devoting their precious time and providing their valuable contributions to this book. I would also like to express my gratitude to my fellow colleagues who encouraged me throughout the process.

Kayden Robertson

1

Introduction to Laser and Laser Science

A laser is a device, which emits amplified light produced as a result of stimulated emission of electromagnetic radiation. The study of lasers is under the domain of laser science. This chapter explores the fundamentals of lasers and laser science.

Laser

The word "laser" is an acronym for the phrase "Light Amplification by the Stimulated Emission of Radiation." A laser is a device in which a collection of atoms or molecules, a semiconductor, or another quantum system, is held between mirrors and energized, or pumped, so that something in the system (the electrons, for example) reach a higher-energy "excited" state.

In a basic laser, a chamber called a *cavity* is designed to internally reflect infrared (IR), visible-light, or ultraviolet (UV) waves so they reinforce each other. The cavity can contain gases, liquids, or solids. The choice of cavity material determines the wavelength of the output. At each end of the cavity, there is a mirror. One mirror is totally reflective, allowing none of the energy to pass through; the other mirror is partially reflective, allowing approximately 5 percent of the energy to pass through. Energy is introduced into the cavity from an external source; this is called *pumping*.

As a result of pumping, an electromagnetic field appears inside the laser cavity at the natural (resonant) frequency of the atoms of the material that fills the cavity. The waves reflect back and forth between the mirrors. The length of the cavity is such that the reflected and re-reflected wave fronts reinforce each other in phase at the natural frequency of the cavity substance. Electromagnetic waves at this resonant frequency emerge from the end of the cavity having the partially-reflective mirror. The output may appear as a continuous beam, or as a series of brief, intense pulses.

The *ruby laser*, a simple and common type, has a rod-shaped cavity made of a mixture of solid aluminum oxide and chromium. The output is in pulses that last approximately 500 microseconds each. Pumping is done by means of a helical flash tube wrapped around the rod. The output is in the red visible range.

A blue laser has a shorter wavelength than the red laser, and the ability to store and read two to four times the amount of data.

The *helium-neon laser* is another popular type, favored by electronics hobbyists because of its moderate cost. As its name implies, it has a cavity filled with helium and neon gases. The output of the device is bright crimson. Other gases can be used instead of helium and neon, producing beams of different wavelengths. Argon produces a laser with blue visible output. A mixture of nitrogen, carbon dioxide, and helium produces IR output.

Lasers are one of the most significant inventions developed during the20th century. They have found a tremendous variety of uses in electronics, computer hardware, medicine, and experimental science.

Laser Components

A generalized laser consists of a lasing medium, a "pumping" system and an optical cavity. The laser material must have a metastable state in which the atoms or molecules can be trapped after receiving energy from the pumping system. Each of these laser components are discussed below.

Pumping Systems

The pumping system imparts energy to the atoms or molecules of the lasing medium enabling them to be raised to an excited "metastable state" creating a population inversion. Optical pumping uses photons provided by a source such as a Xenon gas flash lamp or another laser to transfer energy to the lasing material. The optical source must provide photons which correspond to the allowed transition levels of the lasing material.

Collision pumping relies on the transfer of energy to the lasing material by collision with the atoms (or molecules) of the lasing material. Again, energies which correspond to the allowed transitions must be provided. This is often done by electrical discharge in a pure gas or gas mixture in a tube.

Chemical pumping systems use the binding energy released in chemical reactions to state.

Optical Cavity

An optical cavity is required to provide the amplification desired in the laser and to select the photons which are traveling in the desired direction. As the first atom or molecule in the metastable state of the inverted population decays, it triggers via stimulated emission, the decay of another atom or molecule in the metastable state. If the photons are traveling in a direction which leads to the walls of the lasing material, which is usually in the form of a rod or tube, they are lost and the amplification process terminates. They may actually be reflected at the wall of the rod or tube, but sooner or later they will be lost in the system and will not contribute to the beam.

If, on the other hand, one of the decaying atoms or molecules releases a photon parallel to the axis of the lasing material, it can trigger the emission of another photon and both will be reflected by the mirror on the end of the lasing rod or tube. The reflected photons then pass back through the material triggering further emissions along exactly the same path which are reflected by the mirrors on the ends of the lasing material. As this amplification process continues, a portion of the radiation will always escape through the partially reflecting mirror. When the amount of amplification or gain through this process exceeds the losses in the cavity, laser oscillation is said to occur. In this way, a narrow concentrated beam of coherent light is formed.

The mirrors on the laser optical cavity must be precisely aligned for light beams parallel to the axis. The optical cavity itself, i.e., the lasing medium material must not be a strong absorber of the light energy.

Laser Media

Lasers are commonly designated by the type of lasing material employed. There are four types which are: solid state, gas, dye, and semiconductor. The characteristics of each type will be described.

Solid state lasers employ a lasing material distributed in a solid matrix. One example is the neodymium: YAG laser (Nd: YAG). The term: YAG is an abbreviation for the crystal: Yttrium Aluminum Garnet which serves as the host for the Neodymium ions. This laser emits an infrared beam at the wavelength of 1.064 µm (µm = 10(-6) meters). Accessory devices that may be internal or external to the cavity may be used to convert the output to visible or ultraviolet wavelength.

Gas lasers use a gas or a mixture of gases within a tube. The most common gas laser uses a mixture of helium and neon (HeNe), with a primary output of 632.8 nm (nm = 10(-9) meter) which is a visible red color. It was first developed in 1961 and has proved to be the forerunner of a whole family of gas lasers. All gas lasers are quite similar in construction and behavior. For example, the carbon dioxide (CO_2) gas laser radiates at 10.6 µm in the far-infrared spectrum. Argon and krypton gas lasers operate with multiple frequency emissions principally in the visible spectra. The main emission wavelengths of an argon laser are 488 and 514 nm.

Dye lasers use a laser medium that is usually a complex organic dye in liquid solution or suspension. The most striking feature of these lasers is their "tunability."

Proper choice of the dye and its concentration allows the production of laser light over a broad range of wavelengths in or near the visible spectrum.

Dye lasers commonly employ optical pumping although some types have used chemical reaction pumping. The most commonly used dye is Rhodamine 6G which provides tunability over 200 nm bandwidth in the red portion (620 nm) of the spectrum.

Semiconductor lasers (sometimes referred to as diode lasers) are not to be confused with solid state lasers. Semiconductor devices consist of two layers of semiconductor material sandwiched together. These lasers are generally very small physically, and individually of only modest power. However, they may be built into larger arrays. The most common diode laser is the Gallium Arsenide diode laser with a central emission of 840 nm.

Time Modes of Operation

The different time modes of operation of a laser are distinguished by the rate at which energy is delivered:

- Continuous Wave (CW) lasers operate with a stable average beam power. In most higher power systems, one is able to adjust the power. In low power gas lasers, such as HeNe, the power level is fixed by design and performance usually degrades with long term use.
- Single Pulsed (normal mode) lasers generally have pulse durations of a few hundred microseconds to a few milliseconds. This mode of operation is sometimes referred to as long pulse or normal mode.
- Single Pulsed Q-Switched lasers are the result of an intra-cavity delay (Q-switch cell) which allows the laser media to store a maximum of potential energy. Then, under optimum gain conditions, emission occurs in single pulses; typically of 10(-8) second time domain. These pulses will have high peak powers often in the range from 10(6) to 10(9) Watts peak.
- Repetitively Pulsed or scanning lasers generally involve the operation of pulsed laser performance operating at a fixed (or variable) pulse rates which may range from a few pulses per second to as high as 20,000 pulses per second. The direction of a CW laser can be scanned rapidly using optical scanning systems to produce the equivalent of a repetitively pulsed output at a given location.
- Mode Locked lasers operate as a result of the resonant modes of the optical cavity which can effect the characteristics of the output beam. When the phases of different frequency modes are synchronized, i.e., "locked together," the different modes will interfere with one another to generate a beat effect. The result is a laser output which is observed as regularly spaced pulsations. Lasers operating in this mode-locked fashion, usually produce a train of regularly spaced pulses, each having a duration of 10(-15) (femto) to 10(-12) (pico) sec. A mode-locked laser can deliver extremely high peak powers than the same laser operating in the Q-switched mode. These pulses will have enormous peak powers often in the range from 10(12) Watts peak.

Laser Working

A laser oscillator usually comprises an optical resonator (laser resonator, laser cavity) in which light can circulate (e.g. between two mirrors), and within this resonator a gain medium (e.g. a laser crystal), which serves to amplify the light. Without the gain medium, the circulating light would become weaker and weaker in each resonator round trip, because it experiences some losses, e.g. upon reflection at mirrors. However, the gain medium can amplify the circulating light, thus compensating the losses if the gain is high enough. The gain medium requires some external supply of energy – it needs to be "pumped", e.g. by injecting light (*optical pumping*) or an electric current (*electrical pumping → semiconductor lasers*). The principle of laser amplification is stimulated emission.

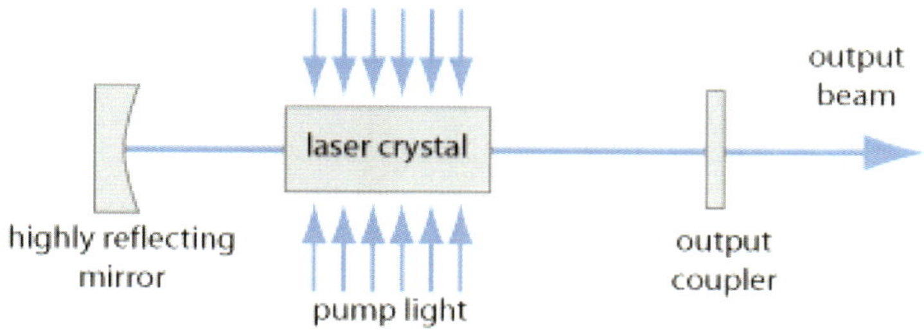

Fig: Setup of a simple optically pumped laser

A laser cannot operate if the gain is smaller than the resonator losses; the device is then below the so-called laser threshold and only emits some luminescence light. Significant power output is achieved only for pump powers above the laser threshold, where the gain can exceed the resonator losses.

If the gain is larger than the losses, the power of the light in the laser resonator quickly rises, starting e.g. with low levels of light from fluorescence. As high laser powers saturate the gain, the laser power will in the steady state reach a level so that the saturated gain just equals the resonator losses (→ *gain clamping*). Before reaching this steady state, a laser usually undergoes some relaxation oscillations. The threshold pump power is the pump power where the small-signal gain is just sufficient for lasing.

Some fraction of the light power circulating in the resonator is usually transmitted by a partially transparent mirror, the so-called output coupler mirror. The resulting beam constitutes the useful output of the laser. The transmission of the output coupler mirror can be optimized for maximum output power.

Some lasers are operated in a continuous fashion, whereas others generate pulses, which can be particularly intense. There are various methods for pulse generation with lasers, allowing the generation of pulses with durations of microseconds, nanoseconds, picoseconds, or even down a few femtoseconds (→ *ultra short pulses* from *mode-locked lasers*).

The optical bandwidth (or line width) of a continuously operating laser may be very small when only a single resonator mode can oscillate (→ *single-frequency operation*). In other cases, particularly for mode-locked lasers, the bandwidth can be very large – in extreme cases, it can span about a full octave. The center frequency of the laser radiation is typically near the frequency of maximum gain, but if the resonator losses are made frequency-dependent, the laser wavelength can be tuned within the range where sufficient gain is available. Some broadband gain media such as Ti:sapphire and Cr:ZnSe allow wavelength tuning over hundreds of nanometers.

Due to various influences, the output of lasers always contains some noise in properties such as the output power or optical phase.

Laser Beam Parameters

The following seven properties are common to the beams emitted from all laser types and are the factors which, when combined together, distinguish laser outputs from other sources of electromagnetic radiation:

1. A nearly single frequency operation of low bandwidth (i.e., an almost pure monochromatic light beam).
2. A beam with a Gaussian beam intensity profile.
3. A beam of small divergence.
4. A beam of enormous intensity.
5. A beam which maintains a high degree of temporal and spatial coherence.
6. A beam that is, in many laser devices, highly plane polarized.
7. A beam with enormous electromagnetic field strengths.

K. Single Frequency Operation (Monochromacity)

The frequency of any electro-magnetic wave is related to the number of cycles the electric or magnetic field undergo each second. A completely coherent, monochromatic wave oscillates exactly at a constant frequency. Most laser systems display a narrow multi frequency characteristic. This frequency spread is, however, very narrow when compared to the average laser frequency.

In most lasers, the frequency degeneracy is solely dependent upon the quantum transition characteristics of the active media, and the geometry of the laser resonator cavity (Fabry-Perot). In this sense, the laser media may be considered as a high number of isolated light generators placed between two mirrors. The electromagnetic field developed between the mirrors may be regarded as a superposition of plane waves at each of the slightly different frequencies which the laser media generates and allows to oscillate. These different frequencies are termed the "modes" of the laser resonator. The off-axis modes result from plane waves propagating at an angle with respect to the axis of the resonator. These different modes are produced by diffraction effects in the Fabry-Perot cavity. The lowest order axial mode is designated as the TEM(oo) mode. This mode has the lowest diffraction losses and often will be the predominant mode of oscillation.

For each transverse mode, there will be many longitudinal modes which can oscillate; hence the output of a multimode laser will actually contain a superposition of plane waves oscillating at many discrete frequencies. However, as previously mentioned, this frequency spread will be very small. In each laser, there will be specific "allowed" frequencies of the resonator cavity (Fabry-Perot modes). For most cases, the average wavelength at which the laser oscillates is sufficient to describe it's operation. If more precision is needed, then the frequency spread or bandwidth is given. Depending on the type of laser, bandwidths range typically from 10(-4) to 10(-9) times the average frequency of the laser; although bandwidths as low as 0.1 Hz. have been reported for stabilized gas lasers.

The wave nature of light most often allows adequate description of the output of the laser, and for most cases it will be sufficient to use geometrical optics to describe the output as a beam with well-defined edges and some beam divergence. The beam is emitted from the laser with a beam diameter (alpha) and a beam divergence (PHI), as though it came from a small point source far behind the laser output aperture.

L. Point Source Emission

The emission from most lasers can be considered as emanating from a "virtual point source" located within or behind the laser device. A "virtual point source" is one which really doesn't exist, but the properties of the emitted beam are such that there appears to be a source at this position. Thus, the virtual point source is located two meters behind the exit mirror.

M. Gaussian Distribution of the Beam

The intensity profile across a TEM(oo) laser beam will be in the form of a bell-shaped (Gaussian) distribution. The decrease in intensity at the edge of the beam is the result of diffraction effects produced at the edges.

The spatial intensity distribution of this mode may be expressed by equation: where R is the radius and W is a constant which defines the mean radius and is commonly referred to as the "spot size." At this point the intensity has fallen to e(-2) of the peak intensity at the center of distribution.

In fact, the edges f the laser beam are not well defined. If one were to measure the energy or power per unit area point by point across the center of the output aperture, a Gaussian beam distribution is defined. The peak intensity is in the center of the beam and approaches zero as one moves from the center. This shape is maintained as the beam propagates through space subject to broadening and distortion by atmospheric effects.

Important points on the distribution curve are the e(-1) and e(-2) intensity points since they are used as standard quantities to define the laser beam divergence parameter. (The e is the natural number associated with the natural logarithm and is equal to: e= 2.7183). The e(-1) point is where the intensity is reduced by the factor or approximately 63% of the energy (or power) is contained within the aperture of diameter (a) centered in the beam.

In most all laser safety and compliance standards, the output aperture, a, and the beam divergence, (PHI), are defined relative to the e(-1) points. The total laser energy (Q(t)) or beam power (PHI) is defined as that which is collected from the entire beam (total values).

Many manufactures specifications use the e(-2) i points to define beam divergence. In this case, e(-2) i = 0.1353, or the total power (energy) is: 100% – 0.13 3 x 100% = 86.47% or approximately 86% of the total energy /power is within the e(-2) aperture. In some cases, they specify relative to the 90% point. Note that the beam divergence is larger at the e(-2) or 90% point. Hazard calculations are sensitive to the beam divergence and conversions from e(-2) points to e(-2) power points are often performed on beam sizes.

The beam diameters at the two points are related. Departure from the Gaussian distribution arise when independent oscillation occurs within the resonator at higher order modes. For example, some gas lasers may be designed to have sufficient gain to support simultaneous oscillation in many different transverse modes. Mode selection may often be accomplished by slight adjustment of the mirror alignments. With this technique, one can observe the different complex intensity distributions of each mode.

The lowest order TEM(00) mode with the nearly Gaussian intensity distribution has the lowest cavity losses and hence will generally be the dominant mode of oscillation.

Optically pumped solid-state lasers such as the normal mode Nd: YAG laser usually display a randomly varying mode output. Thermal gradients in the optical media (i.e., the crystal) caused by non-uniform absorption of the pump light give rise to lens effects in the crystal which change during the pumping cycle. The result is a sporadic switching of transverse modes during the laser pulse. The time average is generally a bell-shaped distribution which is dependent upon the optical purity of the laser crystal, the pumping scheme, and the level at which the system is operated above lasing threshold.

Some pumping schemes produce pronounced "hot spots" in the intensity distributions. For long range transmission, atmospheric effects can also produce intensity variations by a factor of ten over localized regions of the beam. Such non-uniformities distribution make it difficult to specify the cross sectional area of the beam. As a result, an average value of beam radius must be chosen. Typically, this is often (1) the half-power point; (2) the e(-1) power point; or (3) the e(-2) power point.

A more precise laboratory practice is to measure the diameter at the stated power point on a densitometer recording obtained from a photographic negative of the output beam distribution.

In the case of optically pumped solid-state lasers, the size of the beam cross section is generally a function of the pumping level of the laser. In general, the higher the pumping level, the wider the beam size. Only when pulsed lasers are operated near threshold, or in special cavity conditions, will the zero order mode (lowest beam spread) predominate.

N. Beam Divergence

Beam divergence is a very important laser parameter and is often expressed in units of milliradians. The symmetry of the laser beam allows the geometry to be reduced to the two dimensions of a plane. The angle (phi), in radians can be related to degrees by noting that for a full circle, phi is 360 degrees.

For some smaller angle, the arc length(s) intercepted along the circumference of the circle can be used to define the angle as. The minimum beam divergence, called the diffraction limited beam divergence, is related by an equation. This concept is expanded to three dimensions by introducing the concept of solid angle. The solid angle (OMEGA) is expressed in units of steradians (sr) and is determined by using the area cut out of a surface of a sphere divided by the square of the distance to that surface; that is. For a sphere, the solid angle may be opened up to include the entire sphere surface area (A = 4 pi R(2)), therefore, the output of a typical laser will be confined to less than 10(-6) sr.

O. Intensity of Laser Emission

In many applications, the most important laser beam charactertic is the enormous intensity of the beam. Intensity is related to the beam power the cross sectional area and the manner in which the beam spreads from one point in space to the next.

Power, by definition, is the time-rate at which work is done; specifically, it is the rate at which energy is used or produced. Energy relates the ability to do work. As with other forms of energy (eg, chemical, mechanical, electrical), electromagnetic energy (light energy) is a conserved quantity. The intensity of the laser is usually expressed by the irradiance (power/area) of the beam. This is determined by dividing the average value of beam power by the average value of the beam cross section. Irradiance units are expressed in Watts per square centimeter.

In pulsed laser operation, instantaneous (peak) Irradiances in excess of 100,000 W/cm(2) are quite easily generated in an unfocused high energy pulsed solid state laser pulse. If this output were contained within a typical beam divergence of 20 milliradians and focused by only moderate power optics, the Irradiance at the focal plane would be increased at least one-hundred fold.

A CW laser is rated in Watts and a pulsed laser is normally rated according to the total energy (Joules) per pulse. Pulsed outputs are also expressed as a radiant exposure in units of Joules per square centimeter.

In order to determine the peak power of pulsed laser, it is necessary to know the pulse shape and duration. The peak power may be closely approximated by assuming a triangular pulse shape and dividing the energy per pulse by the pulse duration at half power. Average power is an important factor for high PRF lasers when determining the laser classification and maximum permissible exposure levels.

The radiometric units of radiance and integrated radiance are used to describe the diffuse reflection of a continuous wave or pulsed laser beam.

Radiance is expressed, by definition, as the Irradiance per unit solid angle (Watts per square centimeter per steradian).

Integrated Radiance is expressed as the radiant exposure per unit solid angle (Joules per square centimeter per steradian).

The unit of solid angle is defined such that all space about a point source (i.e., the source of light) will encompass 4 pi sr.

P. Focused Laser Beams

The beam from an ideal laser, i.e., a laser which emits a coherent wave, can be considered as a diffraction-limited beam. In this case, divergence of the beam is limited to the effects of diffraction at the beam edges. The emission from such a laser will display a far-field diffraction pattern at a distance where a is the diameter of the emergent laser radiation.

The TEM(00) beam from a typical helium neon laser will display a 0.5-1.0 milliradian beam spread at a distance of 1.0-2.0 meters from the laser.

Due to the high degree of coherence of a laser beam, it is theoretically possible to focus the beam to the diffraction limit of the wavelength of light. Typically, however, the laser will have a finite beam spread and can be expressed by the simple equations of geometrical optics.

The spot diameter (d) is given by the simple equation: d = f phi where: d = spot diameter at focus f = focal length of lens phi = laser beam divergence (radians).

As an example, one can calculate the spot size of a beam focused on the human retina. For this case, consider a "typical" HeNe laser where: phi = 1.0 milliradian and assume that the effective focal length (f) of the human eye is 1.7 cm.

Thus: d = f phi = (1.7 cm) x (1.0 x 10(-3) rad.) = 17 x 10(-4) cm = 17 μm

To give some idea of how small this focused spot is, consider that 17 micrometers is approximately the size of two or three human blood cells stacked end-to-end.

Using the equation for the area of a circle, one can now calculate the focused beam area.

As the spot diameter approaches the wavelength of light, the spot becomes diffraction-limited. For example, the beam from a highly coherent single transverse mode (TEM(00)) gas laser will produce a Gaussian intensity pattern when focused. This distribution may be described mathematically by an equation where it is considered that the beam energy will be contained in a diameter defined at the e(-2) power point.

Therefore, the smallest possible spot size of a focused laser beam will approach the dimensions the wavelength of light which is being focused.

Combining the equations above can yield an expression for the spot area.

Thus, the Irradiance (power per unit area) of a focused laser beam will vary inversely with the square of the focal length of the lens and with the square of the beam divergence angle. Hence, these two factors have dramatic effects on the power distribution at the focal plane of the lens.

Consequently, either a reduction in the focal length of the lens used to focus the beam or a reduction in the beam spread by a factor of ten will produce a one-hundred fold increase in the irradiance at the focal plane of the lens. Simultaneous reduction of both by a factor of ten would increase the Irradiance at the focal plane by a factor of 10(4).

In practice, however, it is usually the beam divergence value that limits the focal spot diameter. This is especially true with pulsed laser systems. To achieve high power outputs, the laser crystal is usually pumped well over threshold; consequently, the beam will contain a conglomerate of high order "off-axis" modes which subsequently increase the beam size.

Typical beam divergence values for gas lasers (helium-neon, argon, etc.) will be about one milliradian, (1 milliradian = 3.44 minutes of arc). Solid-state ruby and neodymium lasers generally have a higher beam spread (1- 30 milliradians), due primarily to the high beam divergence associated with the random multimode operation of such devices.

Q. Scanning Lasers

Some laser applications employ electro-mechanical or electro-optical scanner units to allow a raster-scan capability to the beam. In this way, the beam can be scanned over a large area (such as in a laser print maker) or over a small area (such as a laser UPC label reader) in a repeated geometry..

R. Coherence

The coherency of a laser beam relates to the constancy of the spatial and temporal variations in the radiation wave fronts. A high degree of coherence implies a constant phase different between two points on a series of equal-amplitude wave fronts (spatial coherence), and in a correlation in time between the same points on different wave fronts (temporal coherence). The two coherence terms are a part of the overall four-dimensional coherence function which completely describes the degree of coherency of the beam.

If the laser beam is considered as a plane wave traveling in one direction, it will be spatially coherent due to the perpendicularity of the wave fronts in the direction of propagation. Also, due to the monochromatic nature of the laser light, the beam will be temporally coherent; that is, it will display a fixed-phase relation between a part of the beam emitted at one time and portion emitted at another. Should the wavelength (or frequency) change, then the temporal coherency would degrade.

In 1802, Thomas Young performed his classic double-slit experiment to demonstrate the wave nature of light. Sunlight though one pinhole was allowed to illuminate two closely spaced pinholes. Each pinhole acted as a "new source" for light and the waves from each of the two pinholes interfered with one another so-as-to produce corresponding light and dark regions (or fringes – as they are called) at the observation screen. If light was not a wave, it would travel in a straight line from one pinhole to another to fall at two points on the screen. As a wave, however, it is diffracted and bent about the edges of the pinholes such that each pinhole illuminates the entire screen.

In part, if it were not for the diffraction effects produced as a light wave passes through a finite aperture, the plane wave output of laser could theoretically be focused by a lens (such as the human eye) to a real point with minimal spot diameter. Thus, the image Irradiance, would have an infinite value.

Due to the wave nature of light and the corresponding diffraction effects produced by finite apertures, the image a point source of light (provided by any real optical system) is not actually a point.

Such a distribution has a bright central area surrounded by light and dark rings. When an optical system's resolution is only limited by the diffraction effects, it is said to be diffraction limited. Even in this condition, a point is "spread out" as it is imaged. The more defects and aberrations introduced by the optical system, the more the spreading. Each lens images a point with some spreading and the manner which it behaves is defined as the point spread function.

Since an optical system spreads a point source image, there is a corresponding limit to its resolution or ability to separate two points close together. This is especially important when looking at stars through a telescope. There are at least two criteria used to define the resolution of two points close together. A simple one, and the one most commonly used, is the Rayleigh criterion. This states that when the peak of one Airy disk is over the first dark ring of the other, the points are resolved. This is normally defined in terms of the apparent angle between the two points.

Lasers are often referred to as coherent sources, but in fact, they really only partially coherent. Only absolutely monochromatic or single frequency waves are truly coherent; however, lasers are so close, relative to anything else, that a loose definition may seem justified. The degree to which two waves are coherent determines how well they interfere when brought together at some point in space.

A thorough treatment of the subject of partial coherence is far beyond the scope of this guideline; however, there are a few properties worth discussing. One frequently encounters the terms spatial and temporal coherence. Temporal coherence effects are those which arise from the finiteness of the spectral band. An increase in fringe visibility with a decrease in source size is a measure of the spatial coherence. An important measure of coherence is the coherence length, dL, which can be conceptually related to the duration of an uninterrupted wave train. Even in the beam from an "ideal" laser, there will be random fluctuations in the phase difference of the electromagnetic fields at two separate points on a wave front. The distance between points on the wave front for which the average of this phase different is equal to adians is generally defined as the lateral coherence distance. Recombination of the light samples from points separated by a distance equal to, or less than, this amount can produce interference fringes. The distance is a classical measure of the spatial coherence of a light beam as observed in the famous "double slit" experiment of Young.

The temporal coherence is a measure of the length of time that the beam is truly monochromatic. This may be considered as the time during which the amplitude of the electromagnetic field will remain constant at a given point in space while the phase varies linearly with time. During this time, the beam will travel a length dL = c dT defined as the coherence length (where c = 3 x 10(8) m/sec., the velocity of light). Thus the coherence time is the time required for light to travel the coherence length in the direction of travel of the beam.

By virtue of this argument, it is seen that the frequency bandwidth is actually a measure of temporal coherence. Thus a frequency stabilized HeNe gas laser (d(v) = 3-5 hertz) will have a coherence time of several hundred milliseconds and a corresponding coherence length of 10(5) km.

In contrast to the high spectral purity of gas lasers, the coherence lengths of pulsed ruby lasers are in the order of 15 meters with corresponding coherence times in the order of only 100 nanoseconds.

S. Polarization of the Laser Output

The polarization of most lasers is directly related to the nature of the resonator. For example, many high power gas lasers are built with Brewster's angle windows on both ends of the gas discharge tube. Such windows present virtually no losses to a beam which has a linear polarization component lying in the plane of incident. Hence the output will be linearly polarized in this plane.

In some solid-state crystal lasers, for example, the ruby laser, the output will be linearly polarized. This is a result of the birefringent nature of the crystal in which the slower "ordinary" polarized photons will have a longer time to interact with the excited chromium ions, thereby favoring a polarized output in this plane. This is generally only true for ruby crystals operating near lasing threshold unless Brewster's angles are fabricated on the ends of the crystal. This latter practice is often necessary for very high power Q-switched laser systems.

In diode lasers, linear polarized light is also observed. This may be attributed to the linear symmetry of the junction region.

T. Electrical Field Strength

The electromagnetic theory of light depicts a light wave as having instantaneous electric and magnetic fields which oscillate at the same frequency. The electrical (E) and magnetic (H) fields are fixed at right angles and are mutually perpendicular to the direction of propagation of the wave. Of particular importance in the description of laser beam interactions is the magnitude of the electric field associated with the beam.

From classical considerations (using Maxwell's equations) the electric field (E) in volts per centimeter associated with a light beam in a vacuum (or air) of average power (PHI) in Watts, spread over a cross-sectional area (A) in cm^2.

Prior to lasers, the electric fields associated with commonly occurring light sources were most nominal. For example, the electric field of sunlight occurring at the earth's surface is about. This constitutes an average field spread over all the wavelengths present in the "white light" of the sun.

In contrast, the instantaneous electric field associated with an unfocused "Q- Switched" Nd: YAG laser burst operating at a level of 100 megawatts and confined to 3 mm beam diameter.

Such strong fields are also found elsewhere in nature, as they are at the magnitude of the electrostatic cohesive forces which bind atomic structures. Consequently, when a laser beam with a field of comparable magnitude enters a transparent structure, an instantaneous massive redistribution of the electric system of the material can occur due to the interaction of the fields. At the present, the interaction of these enormous electromagnetic fields is not fully understood, to be sure. The production of free electrons, ionized atoms, and X-rays have been detected in the reaction association with the interaction of high power laser beams.

Types of Lasers

Lasers are classified into 4 types based on the type of laser medium used:

- Solid-state laser
- Gas laser
- Liquid laser
- Semiconductor laser.

Solid-state Laser

A solid-state laser is a laser that uses solid as a laser medium. In these lasers, glass or crystalline materials are used.

Ions are introduced as impurities into host material which can be a glass or crystalline. The process of adding impurities to the substance is called doping. Rare earth elements such as cerium (Ce), erbium (Eu), terbium (Tb) etc are most commonly used as dopants.

Materials such as sapphire (Al_2O_3), neodymium-doped yttrium aluminum garnet (Nd: YAG), Neodymium-doped glass (Nd: glass) and ytterbium-doped glass are used as host materials for laser medium. Out of these, neodymium-doped yttrium aluminum garnet (Nd: YAG) is most commonly used.

The first solid-state laser was a ruby laser. It is still used in some applications. In this laser, a ruby crystal is used as a laser medium.

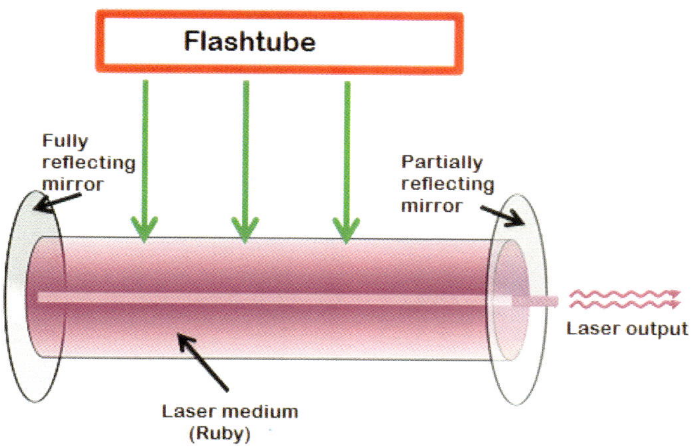

In solid-state lasers, light energy is used as pumping source. Light sources such as flashtube, flash lamps, arc lamps, or laser diodes are used to achieve pumping.

Semiconductor lasers do not belong to this category because these lasers are usually electrically pumped and involve different physical processes.

Gas Laser

A gas laser is a laser in which an electric current is discharged through a gas inside the laser medium to produce laser light. In gas lasers, the laser medium is in the gaseous state.

Gas lasers are used in applications that require laser light with very high beam quality and long coherence lengths.

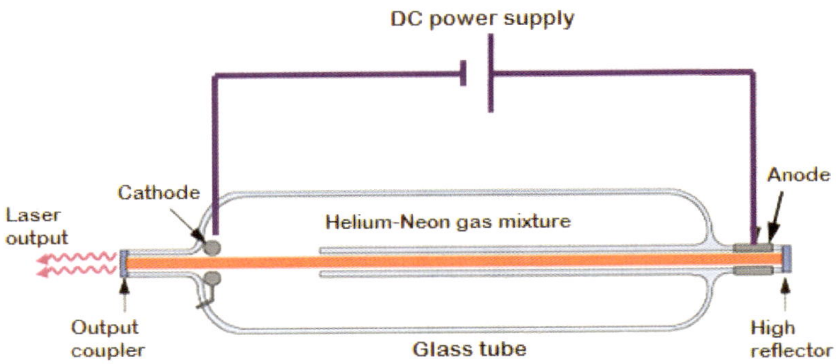

In gas laser, the laser medium or gain medium is made up of the mixture of gases. This mixture is packed up into a glass tube. The glass tube filled with the mixture of gases acts as an active medium or laser medium.

A gas laser is the first laser that works on the principle of converting electrical energy into light energy. It produces a laser light beam in the infrared region of the spectrum at 1.15 μm.

Gas lasers are of different types: they are, Helium (He) – Neon (Ne) lasers, argon ion lasers, carbon dioxide lasers (CO_2 lasers), carbon monoxide lasers (CO lasers), excimer lasers, nitrogen lasers, hydrogen lasers, etc. The type of gas used to construct the laser medium can determine the lasers wavelength or efficiency.

Liquid Laser

A liquid laser is a laser that uses the liquid as laser medium. In liquid lasers, light supplies energy to the laser medium.

A dye laser is an example of the liquid laser. A dye laser is a laser that uses an organic dye (liquid solution) as the laser medium.

A dye laser is made up of an organic dye mixed with a solvent. These lasers generate laser light from the excited energy states of organic dyes dissolved in liquid solvents. It produces laser light beam in the near ultraviolet (UV) to the near infrared (IR) region of the spectrum.

Semiconductor Laser

Semiconductor lasers play an important role in our everyday life. These lasers are very cheap, compact size and consume low power. Semiconductor lasers are also known as laser diodes.

Semiconductor lasers are different from solid-state lasers. In solid-state lasers, light energy is used as the pump source whereas, in semiconductor lasers, electrical energy is used as the pump source.

In semiconductor lasers, a p-n junction of a semiconductor diode forms the active medium or laser medium. The optical gain is produced within the semiconductor material.

A free electron laser is a relatively exotic type of laser where the optical amplification is achieved in an undulator, fed with high energy (relativistic) electrons from an electron accelerator. Such devices have been demonstrated with emission wavelengths reaching from the terahertz region via the far- and near-infrared, the visible and ultraviolet range to the X-ray region, even though no single device can span this whole wavelength range.

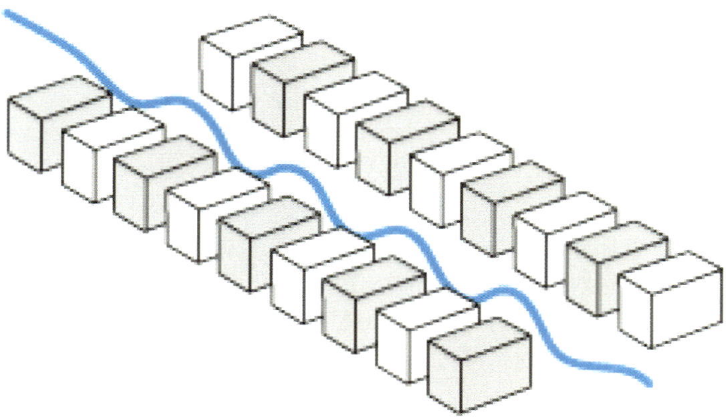

Figure: Setup of an undulator, as used in a free electron laser. The periodically varying magnetic field forces the electron beam (blue) on a slightly oscillatory path, which leads to emission of radiation

In the undulator, a periodic arrangement of magnets (permanent magnets or electromagnets) generates a periodically varying Lorentz force, which forces the electrons to radiate with a frequency which depends on the electron energy, the undulator period, and (weakly) on the magnetic field strength. Both spontaneous and stimulated emission occur, allowing for optical amplification in a certain wavelength range.

The greatest attractions of free electron lasers are:

- Their ability to be operated in very wide wavelength regions
- The large wavelength tuning range possible with a single device
- The spectacular performance in extreme wavelength regions, not reachable with any other light source

Compared with other synchrotron radiation sources (pure undulators and wigglers), FELs can generate an output with a much higher spectral brightness and coherence. This is very useful for a number of applications, including fields such as atomic and molecular physics, ultrafast X-ray science, advanced material studies, ultrafast chemical dynamics, biology and medicine.

The big drawback of FELs is that their setups are very large and expensive, so that they can be used only at relatively few large facilities in the world. The most ambitious free electron laser project is currently pursued in Hamburg (XFEL, within the TESLA project) with the goal of obtaining hard X-ray output with unprecedented performance features: wavelengths down to 0.085 nm and pulse durations below 100 fs. So far, wavelengths down to 6.5 nm have been achieved in the smaller FLASH facility. The LCLS at SLAC has already achieved lasing with 0.15 nm wavelength.

Principles of Operation in FEL

In an FEL a beam of relativistic electrons produced by an electron accelerator passes through a transverse, periodic magnetic field produced by a magnet called an undulator and exchanges energy with an electromagnetic radiation field. Efficient energy exchange requires that the electrons experience nearly resonant forces from the radiation and undulator fields. This resonance is achieved when the radiation wavelength λ, the electron-beam axial velocity $c\beta z$, and the undulator period λ_o satisfy $\lambda \approx \lambda_o(\lambda - \beta_z)$, where c is the speed of light. The resonance relationship shows that the FEL can be continuously tuned by changing the electron-beam kinetic energy $(\gamma - 1)mc^2$, where $\gamma = (\lambda - \beta_\perp^{\,2} - \beta_z^{\,2})^{-1/2}$ is the relativistic Lorentz factor, m is the electron mass, and $c\beta_\perp$ is the transverse electron velocity. It also indicates that the FEL mechanism can be designed to operate over a large range of wavelengths, from centimeters to nanometers. The actual feasibility of operating in a given wavelength range depends on several factors, including the FEL gain and electron-beam quality.

Fig is a schematic of an FEL, with the periodic undulator field indicated by the alternating up and down arrows. The typical peak magnetic field in the undulator is a few kilo gauss with a wavelength of $\lambda_o \approx$ several centimeters over N ≈ 100 periods. The length of the undulator is several meters inside the optical resonator mirrors, which are separated by about twice the undulator length, or about 10 meters. The electron-beam path is indicated horizontally. The beam energy can vary from a few MeV to several GeV, although the typical FEL uses an electron-beam energy of around 50 MeV. The transverse dimensions in the FEL schematic are exaggerated with respect to the longitudinal distance shown. The electron-beam radius is typically about 1 mm, and the optical mode waist is about 3 mm. The electron-beam pulse width can be as long as several microseconds for a recirculating electrostatic accelerator. In a radio frequency (RF) accelerator or storage ring, the electron-beam structure is determined by the RF frequency. In an RF accelerator, only those electrons with the correct phase relation in the RF cycle continue to be accelerated. The resulting accelerated electron current is produced in short micro pulses of a few picoseconds. The minimum separation between micro pulses is one RF cycle, but often there is a much larger separation of many nanoseconds. The stream of micro pulses continues for about 10 μs up to about 1 ms because the RF power source is pulsed. This stream of micro pulses is called a macro pulse. In a storage ring using lower-frequency RF, the micro pulse length is typically a nanosecond in length separated by several RF cycles. There is no macro pulse in a storage ring. The radiation pulse structure in an FEL reflects that of the electron beam.

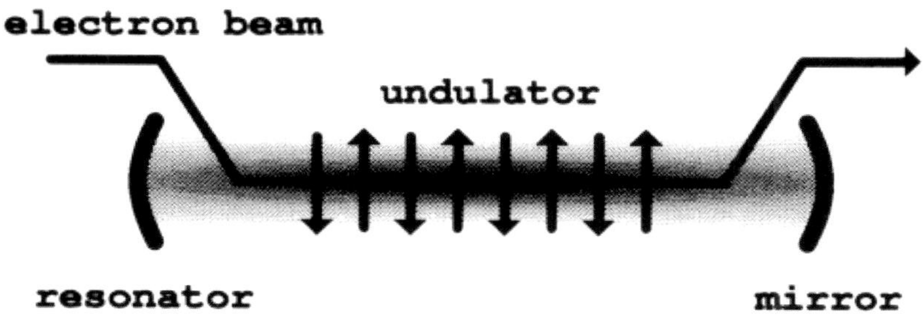

Figure: Schematic of a free electron laser

The requirements of high electron-beam current and quality are crucial in determining the ultimate performance of an FEL. The electron-beam kinetic energy $(\gamma - 1)\, mc^2$ can range from a few MeV to several GeV, while the beam current can range from a few amperes to several kilo amperes. The FEL is able to obtain large peak power, on the order of a few gigawatts, because, unlike conventional lasers or microwave tubes, the energy that is not transferred to the radiation field remains in the relativistic electron beam that is transported out of the undulator at nearly the speed of light, or even recovered to improve overall efficiency. The FEL interaction volume contains only light, the undulator magnetic field, and the electron beam, so that unwanted high field effects and thermal distortion of the medium are absent in the FEL. The small duty cycle of most accelerators limits the average power of FELs. The efficiency of FELs has been demonstrated to be greater than 40% at long wavelengths, but most will operate at a lower efficiency of a few percent.

The classical gain in the low-gain FEL oscillator develops from coherent electron bunching on the scale of the radiation wavelength. As the electrons travel through the undulator, they accelerate from side to side and spontaneously radiate in the forward direction. The spontaneous emission process is the same as is used to generate UV and x-rays in the undulators and wigglers of synchrotron radiation sources. On the first few passes through the FEL oscillator, some of this spontaneous radiation is stored in the resonator. As the radiation power grows over many passes, the emission process begins to differ from that of the synchrotron sources and becomes similar to the mechanism used in the early microwave electron tubes.

The electrons entering the operating FEL are forced to move from side to side due to the undulator magnetic field, but now radiate in the presence of the stored transverse radiation field. As the radiation passes over the more slowly moving electrons, the radiation electric field does work on the oscillating electrons. This is the classical analog of quantum stimulated emission. A significant energy exchange can occur only when the undulator and radiation field forces on the relativistic electrons are nearly resonant. The electrons in the beam are initially randomly spread over many radiation wavelengths with many millions of electrons within each section of the beam one wavelength long. The orientation of the radiation's electric field vector and the electron's undulator velocity vector rotates through 2π within each radiation wavelength. These different orientations determine that about half the electrons lose energy to the radiation field, while other electrons gain energy from the co-moving wave. The electrons that gain energy begin to move ahead of the average electron, while the electrons that lose energy begin to fall behind the average. As the faster electrons move ahead of the average and the slower electrons move back, the beam is periodically bunched on the radiation wavelength scale. Since the electron energy loss and gain repeats throughout the electron beam over many radiation wavelengths, the beam becomes bunched on the scale of the radiation wavelength. The emission rate for a perfectly bunched beam of electrons is proportional to the square of the number of electrons, whereas the emission rate for a beam of randomly positioned electrons is only proportional to the number of electrons. The same kind of electron bunching is the goal of the microwave electron tube design. The FEL works in a similar way, but can reach infrared, optical, UV, or even x-ray wavelengths. The bunched beam then radiates coherently and amplifies the existing radiation.

There is always an energy spread imposed on the electron beam in order to move the electrons together in a bunch. When the beam has an initial energy, or velocity spread, the bunching process can be impaired, reducing gain. In the FEL oscillator working in steady state at high power, the electrons become over-bunched at the end of the undulator and actually take energy back from the radiation field. This is the normal saturation process, and typically limits the fraction of the electron-beam power converted to radiation in a single pass to 1/N.

In an FEL amplifier (as opposed to an oscillator), the high current of the electron beam increases the rate of the electron bunching process along the undulator. For sufficiently high beam current entering a sufficiently long undulator, the same electron bunching process that occurs over many passes in the FEL oscillator can be completed in a single pass along the FEL amplifier. The amplified wavelength can be selected by an external oscillator laser. In the super radiant FEL, there is no external oscillator, so that radiation must first grow from spontaneous emission. As the optical power grows, the spectrum narrows to establish coherence. This process is called self-amplified spontaneous emission (SASE).

The relativistic electrons see each rapidly advancing undulator period Lorentz contracted to a shorter wavelength, $\lambda_o' = \lambda_o/\gamma$. Also, the electrons "see" the radiation field passing over them as Doppler shifted to longer wavelengths, $\lambda' = (1 + \beta_z)\,\gamma\lambda \approx$

2 $\gamma\lambda$. The condition of resonance between the undulator and radiation forces in the beam frame, $\lambda_0' = \lambda'$, gives the FEL resonance condition in the laboratory frame, $\lambda = \lambda_0/2\gamma^2$.

Properties of FEL

Many of the properties of the FEL can be inferred from the above discussion. These properties are summarized below:

- *Tunability:* Because the FEL uses a single gain medium, the relativistic electron beam, and because the resonant condition can be easily tuned by changing either the electron beam energy or the magnetic field strength, FELs are broadly and easily tuned. A factor-of-10 tunable frequency range has already been demonstrated with the same accelerator and undulator.
- *High peak power:* Because waste energy is carried away at nearly the speed of light and because the lasing medium cannot be damaged by high optical fields, FELs can produce very high peak powers. Giga watt peak powers have been demonstrated.
- *Flexible pulse structure:* Because the pulse structure of the radiation follows the pulse structure of the electron beam, the mature RF technology of linear
- accelerators can be used to manipulate and control the FEL pulse structure. Picosecond pulses with sub-picosecond jitter can be produced, the interval between pulses can be varied, and there is the possibility of producing complicated pulse structures.
- Good laser characteristics: FELs easily achieve desirable properties associated with conventional lasers, such as a single transverse mode, high spatial and temporal coherence, and flexible polarization properties.
- Broad wavelength coverage: Because the gain medium is transparent at all wavelengths, FELs in principle can produce radiation at any wavelength. In practice, electron-beam energy, current, emittance, and energy spread requirements become more stringent as the wavelength decreases, and the cost, size, and complexity of the FEL are therefore higher at shorter wavelengths. FELs can have significant emission at harmonics of the fundamental frequency given by the resonance condition, and harmonics can be used to extend operation to shorter wavelengths than would be practical using only the fundamental frequency. At present, the shortest wavelength that has been achieved in an FEL is 240 nm, and significant use of FELs for scientific research has been restricted to the infrared. There are proposals to build vacuum ultraviolet and x-ray FELs.
- Size and cost: Because it requires an electron accelerator with its associated

shielding, the FEL has not been a device that can be placed in an individual investigator's laboratory and be operated and maintained by graduate students whose primary expertise is in other areas of science. FELs have been used principally in central facilities, where their utilization in scientific research involves associated costs of maintaining and operating the facility in addition to the cost of the device itself. FELs have been used by one user, or at most a few users, at any one time, while synchrotrons can be used simultaneously by many users. For example, the APS now under construction at Argonne National Laboratory can have 34 insertion device beamlines and 35 bending magnet beam lines, all of which can be used independently by different users at the same time. The basic science of electron accelerators is well known, but virtually all the effort spent on accelerator physics has been to reach higher energies with bigger and more costly machines. Relatively little effort has been devoted to producing the smaller and less expensive machines that would be most useful for an FEL.

Tunable Laser

A tunable laser is a laser the output wavelength of which can be tuned (i.e. adjusted) (→ *wavelength tuning*). In some cases, a particularly wide *tuning range* is desired, i.e. a wide range of accessible wavelengths, whereas in other cases it is sufficient that the laser wavelength can be tuned (factory-set) to a certain value. Some single-frequency lasers can be continuously tuned over a certain range, whereas others can access only discrete wavelengths or at least exhibit mode hops when being tuned over a larger range. Lasers are sometimes called *wavelength agile* or *frequency agile* when the tuning can be done with high speed.

Tunable lasers are usually operating in a continuous fashion with a small emission bandwidth, although some Q-switched and mode-locked lasers can also be wavelength tuned. In the latter case, it is possible to shift either the envelope of the frequency comb or the lines in the optical spectrum.

Widely Tunable Lasers

Some types of lasers offer particularly broad wavelength tuning ranges:

- A few *solid-state bulk lasers*, in particular titanium–sapphire lasers and Cr:ZnSe and Cr:ZnS lasers allow tuning over hundreds of nanometers in the near- and mid-infrared spectral region. (In general, transition-metal doped gain media offer larger tuning ranges than rare-earth-doped gain media, since the electrons involved in such media interact more strongly with the host lattice.
- Dye lasers also allow for broadband tunability. Different dyes can cover very broad wavelength ranges, e.g. throughout the visible region. There are narrow-line width dye laser systems (continuous-wave or pulsed) for use in spectroscopy, and also mode-locked dye lasers generating femtosecond pulses.

- Some free electron lasers can cover enormously broad wavelength ranges, and often in extreme spectral regions.

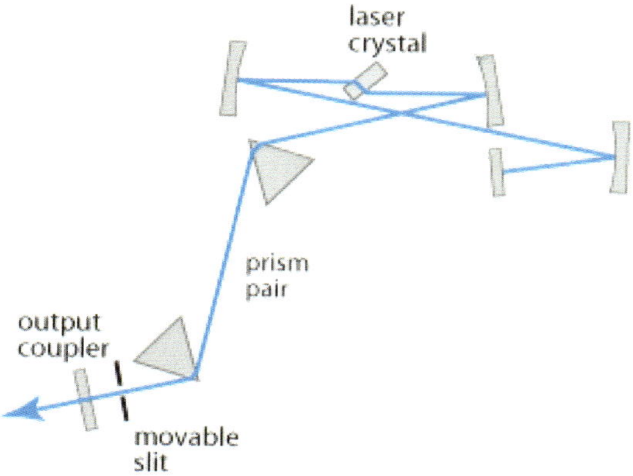

Figure: Setup of a tunable solid-state bulk laser, realized e.g. with a Ti:sapphire laser crystal. The prism pair spatially disperses the different wavelength components, so that the movable slit can be used to shift the wavelength away from that of maximum gain

Other types of lasers offer tuning ranges spanning a few nanometers to some tens of nanometers:

- *Rare-earth-doped fiber lasers*, e.g. based on ytterbium, can often be tuned over tens of nanometers, sometimes even more than 100 nm. Most Raman fiber lasers also have the potential for wideband tuning.
- Some *rare-earth-doped laser crystals*, often doped with ytterbium, also allow for substantial tuning ranges of bulk lasers. Examples are tungstates, vanadates, Yb:BOYS, and Yb:CALGO.
- Color center lasers rely on broadband gain from certain lattice defects in a crystal, which can be generated e.g. with gamma irradiation. They are not widely used, however.
- Most laser diodes can be tuned over a few nanometers (often by varying the junction temperature), but some special types such as external-cavity diode lasers and distributed Bragg reflector lasers can be tuned over 40 nm and more.
- Quantum cascade lasers are also broadly tunable mid-infrared laser sources.

Some fine tuning, often continuously without mode hops, is possible for other lasers:

- Some compact *solid-state bulk lasers* such as non-planar ring oscillators (NPROs, MISERs) allow continuous tuning within their free spectral range of several gigahertz. Tuning may be accomplished by applying stress to the laser crystal via a piezo, or by varying the crystal temperature.
- Similar fine tuning is possible with some *single-frequency laser diodes*, e.g. by varying the drive current.

For wideband tuning in various spectral regions, *optical parametric oscillators* (OPOs) can be used. These are actually not lasers, but OPO sources are nevertheless sometimes included with the term *tunable laser sources.*

Applications of Tunable Lasers

Wavelength-tunable laser sources have many applications, some examples of which are:

- In spectroscopy, a wavelength-tunable laser with narrow optical bandwidth can be used for recording transmission or absorption spectra with very high frequency resolution. In a LIDAR system, a laser may be tuned to a wavelength which is specific to a certain substance to be monitored.
- Various methods of laser cooling require a laser wavelength to be adjusted very precisely at or near some atomic resonance.
- Tuning to atomic resonances is also used in laser isotope separation. The laser is then tuned to a particular isotope in order to ionize these atoms and subsequently deflect them with an electric field.
- A tunable laser can be used for device characterization, e.g. of photonic integrated circuits.
- In optical fiber communications with wavelength division multiplexing, a tunable laser can serve as a spare in the case that one of the fixed-wavelength lasers for the particular channels fails. Even though the cost for a tunable laser is higher, its use can be economical as a single spare laser can work on any transmission channel where it is needed. As the cost of tunable lasers is no longer much higher than for non-tunable ones, tunable lasers are now often even used throughout.
- In optical frequency metrology, it is often necessary to stabilize the wavelength of a laser to a certain reference standard (e.g. a multi pass gas cell or an optical reference cavity). This can be accomplished e.g. with an electronic feedback system, which automatically adjusts the laser wavelength.
- Some interferometers and fiber-optic sensors profit from a wavelength-tunable laser source, e.g. if this makes it possible to remove an ambiguity or to avoid mechanical scanning of an optical path length.

Laser Science

Laser physics is a branch of physics where the fundamental and the applied aspects of laser science are studied. Basically laser physics is a part of optics which deals with the laser theory.

The spatial coherence helps the laser beam to stay narrow over a vast region. This narrowness is helpful for the laser pointer. In this case, the laser beam achieves a very high irradiance and the divergence of light is very low.

Again, laser is also having a high temporal coherence which helps laser beam to give a very narrow spectrum. This temporal coherence laser beam is polarized one.

Laser physics is used to manipulate the single quantum particles for implementation quantum information processing.

To analysis the interaction between light and the matters, laser physics is useful. The high intensity light matter interaction can be measured by the ultrafast optical laser.

Again, to create an exotic quantum state like Bose Einstein condensation state, laser is very useful. Laser physics is also using to determine the quantum mechanical superposition and quantum entanglement information.

In atomics physics, there is also much application of laser physics, like the interaction between the atoms and the photons at the single particle level can be determined.

There are many applications of laser physics in various technologies, like fiber optics technology, optical disk drives, laser printers, barcode scanner, DNA sequencing instrument, laser surgery, skin treatment, nonlinear optics, ultra short pulses, optical communication, general optics, optoelectronics, and quantum optics.

References

- Laser-types-and-operation, laser-safety: cemarelectro.com, Retrieved 22 May 2018
- What-laser: nist.gov, Retrieved 30 June 2018
- Free-electron-lasers: rp-photonics.com, Retrieved 02 March 2018
- Tunable-lasers: rp-photonics.com, Retrieved 29 April 2018
- laser-physics-204: chegg.com, Retrieved 29 July 2018

2

Chemical Lasers

Chemical laser is a laser that is powered by a chemical reaction. These are useful in industrial operations of cutting and drilling. This chapter has been carefully written to provide an easy understanding of the varied aspects of chemical lasers. It includes topics like chemical oxygen iodine laser, all gas-phase iodine laser, HF (DF) laser, MIRACL, etc.

The chemical laser is a type of laser system that is driven by the release of energy during a chemical reaction. Chemical lasers are usually large, high power devices that integrate chemical delivery systems, a supersonic flow apparatus, and an optical resonator.

Polanyi first proposed a chemical laser in 1960 by recognizing that many chemical reactions deposit excess energy in the stretching of the newly formed bond:

1. If the chemical reaction favors the production of states with greater internal energy over states with less internal energy, then a population inversion can be achieved.

   As an example of using a chemical reaction to pump a laser, consider the reaction of atomic fluorine with molecular hydrogen to produce vibrationally excited hydrogen fluoride, as shown in figure below. This reaction produces 130 kJ/mole excess energy which can be deposited in vibration of the newly formed HF bond.

2. The first four vibrational levels, v=0-3, are accessible. The barrier to this reaction is small, 5 kJ/mole, and thus the rate for producing HF is quite rapid. The key to developing an HF laser based on this reaction is the relative rates for producing the various vibrational levels of HF. These relative rates have been measured as 12:20:6:1 for v=3:2:1:0.

3. Thus, 31% of the HF products will be found in the vibrational level v=3. The most probable product is v=2 at 51%. Only 3% of the reactions yield the lowest vibrational state, v=0. Clearly, a population inversion between vibrational levels in HF can be generated by this direct chemical reaction.

In 1965 Kasper and Pimentel demonstrated the first chemical laser by initiating a hydrogen-chlorine explosion with a flash lamp.

(4) Within a few years, purely chemical, continuous operation HF lasers based on the chemistry described above had been demonstrated and by 1984 HF lasers with powers greater than 1 MW (one million watts) were developed.

(5) The HF chemical laser operates in the infrared near a wavelength of 2.7 microns.

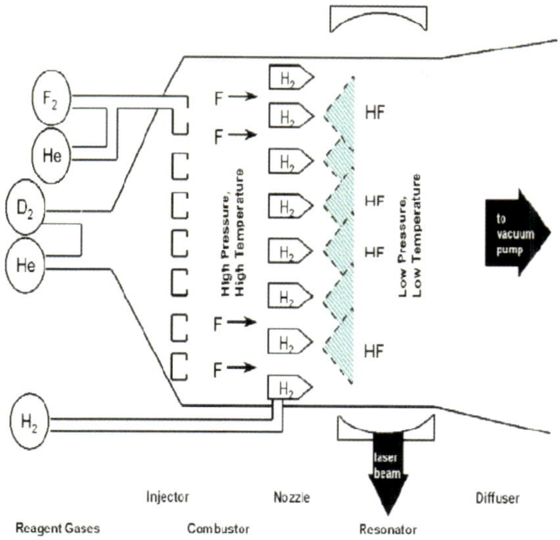

Figure: A typical supersonic mixing hydrogen fluoride chemical laser

The basic components of a combustion driven, supersonic mixing HF laser are shown in figure below. Fluorine atoms must be produced to drive the $F + H_2 + \rightarrow HF + H$ reaction and this is accomplished by thermal dissociation of molecular fluorine at high pressure and temperature in the combustor. Typically, molecular deuterium is injected as the fuel and burned with molecular fluorine or NF_3 to produce 30-50% dissociation of F_2 + at 2000 - 4000 ^{O}K and 1 - 15 kPa total pressure. These temperature and pressure conditions are far too high for efficient laser operation, and the effluents of the combustor including the atomic fluorine are supersonically expanded through a nozzle assembly at Mach 1-3 producing a temperature of 300-500 ^{O}K and a pressure of 0.2 - 0.8 kPa in the gain region. While the flow is cooled, the fluorine atoms remain dissociated as there is insufficient time for recombination to molecular fluorine. Molecular hydrogen is then injected into this supersonic expansion through a large number of very small nozzles to enable good mixing and efficient reaction with the atomic fluorine to produce the vibrationally inverted HF molecules for lasing. Another function of the supersonic flow is stretch the reaction zone for the $F + H_2 \rightarrow HF + H$ reaction over the full width of the laser resonator. The laser cavity is formed so that the laser beam propagates in a direction perpendicular to the gas flow. Finally, the diffuser is designed to recovery as much of the inlet stagnation pressure as possible before exiting to the mechanical or steam injection vacuum pumping system where the gases are exhausted.

The DF chemical laser substitutes deuterium for hydrogen to react with the atomic fluorine, $F + D_2 \rightarrow DF + D$, and uses hardware very similar to the HF laser. The vibrational energy for DF is less than HF, and the DF laser operates at a wavelength near 3.8 microns. The atmosphere exhibits a window with good transmission at 3-5 microns, making the DF laser preferred for applications requiring atmospheric propagation.

Chemical Laser Performance Parameters			
Laser Parameter	Symbol	Typical Value	
		HF Laser	COIL
Wavelength	λ	2.7 μm	1.3 μm
Gain	γ	20 % /cm	1 % / cm
Mass Efficiency	σ_m	150 kJ/kg	300 kJ/kg
Nozzle Flux	δ	---	200 kW/cm^2
Beam Quality	BQ	2	1.3
Saturation Intensity	I_{sat}	0.1 kW/cm^2	5 kW/cm^2

Chemical Laser Performance

Several key parameters that describe the performance of a chemical laser are defined in Table 1. The laser gain, γ, is specified by the product of the cross-section for stimulated emission, σ ,and the population inversion, Δ:

$$\gamma = \sigma\,\Delta = \sigma\left[N_2 - (g_2/g_1)N_1\right]$$

Where, N_2 and N_1 are the concentrations and g_2 and g_1 are the degeneracies for the upper and lower laser levels, respectively. The upper and lower laser levels are labeled by their energies, E_2 and E_1. The cross-section for stimulated emission is related to the spontaneous emission rate:

$$\sigma = A_{21}\left(\lambda_0^2/8\pi\right)g(\lambda)$$

Where,

A_{21} = rate for spontaneous emission from level E_2 to level $E_2\left(s^{-1}\right)$

$\lambda_0 = hc/(E_2 - E_1) =$ laser wavelength (nm)

$g(\lambda) =$ line shape function specifying the shape and width of the spontaneous emission centered at wavelength $\lambda_0\ (s)$

Typically, the total pressure in the gain region of a chemical laser is low and the transition is inhomogeneously broadened with a Gaussian line shape function:

$$g(\lambda) = \sqrt{\frac{4\,ln}{\pi}}\left(\frac{1}{\Delta v_D}\right)\exp\left\{-4\,ln\,2\left(\frac{c/\lambda - c/\lambda_0}{\Delta v_D}\right)^2\right\}$$

Where

$$\Delta v_D = \sqrt{\frac{8kT\,ln\,2}{Mc^2}}\left(\frac{c}{\lambda_0}\right) = \text{Doppler line width (Hz)}$$

$T =$ gas temperature $\left({}^{o}K\right)$

M = mass per emitter particle (kg)

When the gain exceeds the total optical losses, α, due to mirror output coupling, scattering losses, and resonator losses, then net amplification and lasing may be achieved. The threshold population inversion is defined as the inversion required for the gain to equal the losses, $\gamma_{th} = \alpha$, or

$$\Delta_{th} = \alpha / \sigma$$

The HF laser exhibits high gain which enables the design of optical resonators with large output coupling. However, large gain also limits the length of the cavity, as large amplification can be established without many round trips in the cavity. Such a situation leads to amplified spontaneous emission without good beam quality and coherence properties. Low gain lasers operate nearer threshold and optical losses more significantly diminish efficiency, but allow for longer gain media and greater geometrical scaling for high power performance.

The inversion, Δ, is established via the pumping process and the time dependence of the population inversion is described by the generic laser rate equations:

$$\frac{dN_2}{dt} = R_{p2} - \Gamma_2 N_2 - \sigma\left(N_2 - \frac{g_2}{g_1}N_1\right)(I_\nu / h\nu)$$

$$\frac{dN_1}{dt} = R_{p1} + \Gamma_{21} + \sigma\left(N_2 - \frac{g_2}{g_1}N_1\right)(I_\nu / h\nu) - \Gamma_1$$

$$\frac{dI_\nu}{dt} = A_{21}N_2\left(\frac{V g(\lambda)}{V_m \rho(\lambda)}\right) + c\sigma\left(N_2 - \frac{g_2}{g_1}N_1\right)(I_\nu) - \alpha c I_\nu$$

The rate, R_{p2}, for chemical production of the upper laser level will depend on the concentration of reagents and is specific to the chemical reaction employed. The chemical pumping reactions may also generate some population in the lower laser level.

The rate for this undesirable process is R_{p1}. The decay rates, Γ_1 and Γ_2, include all radiative and collisional processes that remove population from the upper and lower laser levels, respectively. Any process that moves population from the upper laser level to the lower laser level is represented by the decay rate, Γ_{21}. The intensity of laser radiation inside the laser cavity, I_n, grows initially from the spontaneous emission into a specific cavity mode with volume, V_m, represented by the first term in the equation for dI_υ / dt. The mode density is $\rho(\lambda) = 8\pi / c\lambda^2$. The second and third terms in the laser intensity equation represent laser amplification due to stimulated emission and loss of photons from the laser cavity, respectively.

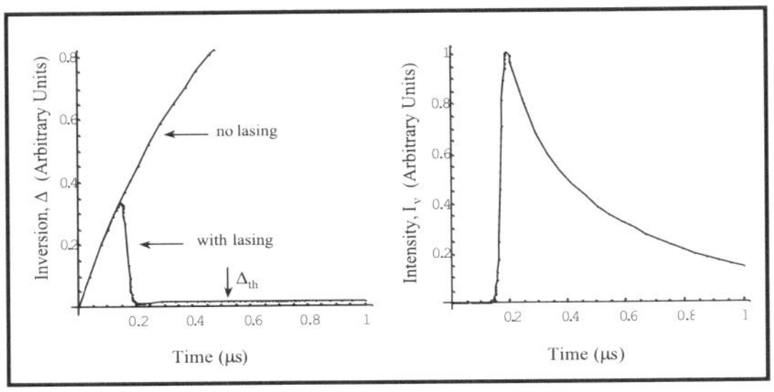

Figure: Solution of the laser rate equations for the inversion and laser intensity for a high gain, low optical loss system where chemical pumping of the upper laser level is initiated at t=0

An example solution to the laser rate equations is shown in figure above. The chemical pumping rate begins suddenly at t = 0 and the population inversion increases rapidly. The inversion reaches threshold almost immediately, as the losses for this system are low. The population inversion, $\Delta = N_2 - N_1$ for two cases: (1) no stimulated emission, (2) with lasing, are both shown. The lasing is delayed by t = 0.2 ms as the photons circulate between the mirrors until a large pulse is achieved. During lasing, the inversion is rapidly depleted by stimulated emission and maintained at the threshold value. The intensity of the laser pulse decays at longer times as initial reagents are consumed and population in the lower laser level increases. Most chemical lasers are continuous-wave (cw) or steady-state devices where reagents are continually replenished in a flowing system. Under such conditions the laser intensity quickly achieves a steady value.

The requirements for pump rates of a chemical laser can be understood from a steady state analysis of the rate equations above. For a molecular system with many accessible rotational levels, pump rates of 10^{19} molecules/cm^3 -s are required to achieve a gain of greater than 0.1%/cm in the visible portion of the spectrum, $\lambda \sim 500$ nm. For a bimolecular pumping reaction with a nearly gas kinetic rate coefficient, the product of reagent concentrations required to achieve this pump rate is approximately 10^{29} molecules 2/cm^6. These rapid excitation rates are often limited by diffusion of reagents in a supersonic mixing nozzle. If the time scale for mixing is long compared to the radiative lifetime, then the effective excitation rate is reduced and the required reagent concentrations are increased. Current nozzle technology limits the mixing time to values greater than 1-10 μs.

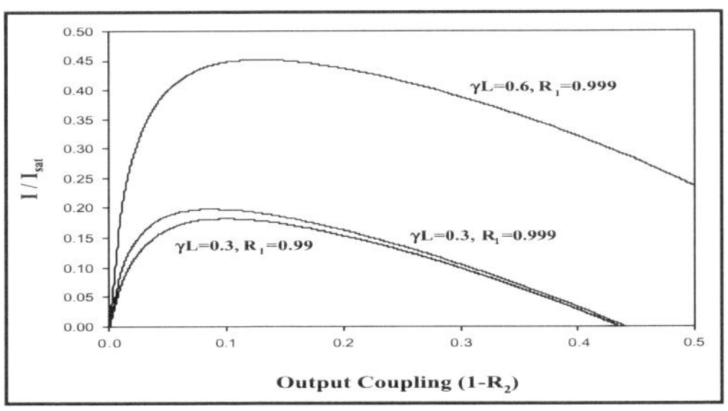

Figure: A Rigrod analysis of the laser intensity as a function of output mirror reflectivity, R_2, for several values of gain and optical losses at the second mirror with maximum reflectivity, R_1

The power output from the laser depends on many factors including the energy density created by the pump reaction, laser gain and losses, output coupling, and the saturation intensity. For example, the output power as a function of mirror reflectivity is shown in Figure 6 for a system with low gain and high saturation intensity. The saturation intensity is dependent only on parameters of the gain media:

$$I_{sat} = (hc/\lambda)\ /\ (\sigma/\Gamma 2)$$

Clearly there is an optimum output coupling (mirror reflectivity) to achieve maximum laser output power which depends on gain, losses, and saturation intensity.

Two key laser performance parameters describing the output power are mass efficiency, σ_m (kJ/kg), and nozzle flux, δ (kW/cm^2). The mass efficiency describes the laser power achieved per reagent flow rate and is particularly important for space based applications where the cost of delivering fuel to orbit can dominate total system costs. Typical mass efficiencies for HF lasers are ~150 kJ/kg and for COIL devices, ~300 kJ/kg. By comparison dynamite (TNT) possesses. an energy density of 9 MJ/kg. The nozzle flux parameter represents the laser power achieved per unit cross-sectional area of the nozzle assembly. This is a key parameter for power scaling and specifies the size of a high power chemical laser.

Beam quality is a key parameter which describes the ability to propagate the laser as a narrow beam and to focus the beam to a small spot. The angular divergence of a laser beam, $\theta_{1/2}$, is ultimately limited by diffraction, $\theta_{1/2} \sim (\lambda/D)$, where D is the diameter of the limiting aperture, or cavity mirror. However, such a diffraction limited beam is not achieved, and the beam quality (BQ) is often reported as the ratio of the actual spot size to the diffraction limited spot size. Chemical lasers, particularly the COIL device, have very good beam quality, due to the transport of excess heat from the laser cavity via the supersonic flow. Beam quality for a chemical laser is typically limited by the spatial uniformity of the gain media imposed by the supersonic mixing nozzles.

The ability of a laser to damage a distant target is often characterized by the source brightness, or power per unit solid angle of beam divergence. The brightness, B, for a diffraction limited beam is aproximately related to laser power, P, operating wavelength, λ, and diameter of the laser aperature, D:

$$B = P\, D^2 / \pi\, \lambda^2$$

It is the power flux, or power per unit area that causes damage due to rapid heating, and the area to which a laser beam can be focused is limited by diffraction, beam quality, and beam jitter. Device powers of multi-megawatts and bright-nesses of 10^{20} - 10^{23} Watts/steradian are required for some military missions.

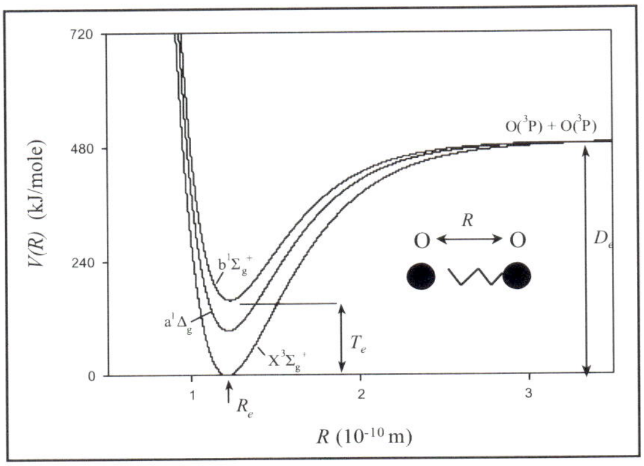

Figure: Morse potential energy curves for the three lowest electronic states of molecular oxygen, which dissociate to ground state oxygen atoms, $O(^3P)$ The dissociation energy, D_e, for the ground state, $X^3\sum_g^+$, and the electronic energy, T_e for the second excited state, $b^1\sum_g^+$, are indicated

The Metastable Energy Transfer Chemical Laser

To extend chemical laser operation to significantly shorter wavelengths than the HF/DF laser requires a system which establishes an inversion between the more energetic electronic states of an atom or molecule. There are several advantages to shorter wavelength chemical lasers including higher power per mass of reagents and reduced beam divergence due to diffraction at the limiting aperture. Rather than depositing excess reaction energy into the stretching of the newly formed bond (vibration), the chemical reaction could yield atoms or molecules with internal excitation of the electrons. For example, molecular oxygen has two excited electronic states with electronic energy of T_e = 95 and 158 kJ/mole above the ground electronic state, as shown in the potential energy curves of figure. To understand these curves, consider the oxygen molecule as two masses connected by a spring, where the bonding electrons shared by the two oxygen atoms provide the restoring force. As the oxygen molecule vibrates, the distance between the two oxygen atoms, R, changes between the two extremes, and the potential energy curve represents this range in inter-nuclear separation at various energies, V. At large enough vibrational energy, the internuclear separation becomes quite large and the molecule dissociates to two ground state oxygen atoms. The curve labeled $X^3\sum_g^+$, is the lowest electronic state of molecular oxygen, which has a dissociation energy of De = 489 kJ/mole and a bond length of Re= 0.1207 nm. If the electron cloud around the oxygen molecule is excited, then the effective spring force constant is changed, and the excited electronic states, labeled $a^1\Delta_g$ and $b^1\sum_g^+$, are generated.

To efficiently produce an inversion between electronic states requires the chemical reaction to selectively produce a specific excited electronic state over all other states, particularly the ground electronic state. Reactions involving light atoms and molecules tend to preserve the total electronic spin. Thus, the generation of an electronic inversion is aided by a difference in electronic spin between the ground and excited electronic states. However, such transitions exhibit slow spontaneous emission rates, and thus very low stimulated emission cross-sections and insufficient gain to build a laser. For example, the $a^1\Delta - X^3\sum$ transition in O_2 shown in figure exhibits a particularly slow spontaneous emission rate of 2.58 x 10-4 s-1. However, these long-lived, or metastable, electronically excited states establish a large energy reservoir that could be used to pump a chemical laser. By collisionally transferring the energy stored in these metastable states to a suitable lasant specie, a new class of chemical laser, the metastable energy transfer chemical laser, can be developed. The chemical oxygen-Iodine is the prime example of such a energy transfer laser.

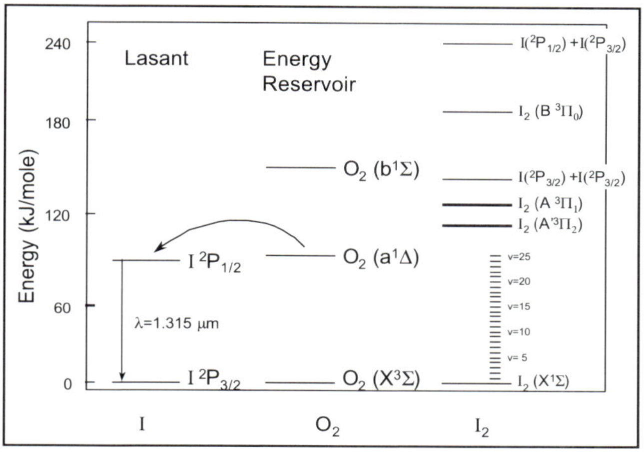

Figure: Energy level diagram for the Chemical Oxygen-Iodine Laser illustrating the resonant energy transfer from metastable reservoir $O_2\left(a^1\Delta\right)$ to the upper laser level of atomic iodine, $I(^2P_{1/2})$

Chemical Oxygen Iodine Laser

Chemical oxygen iodine laser (COIL) is an infrared chemical laser that has been developed at the Air Force Research Laboratory in 1977, primarily for military applications. It outputs a beam of infrared at a wavelength of 1.315 μm. In continuous mode, the output power of the laser can be scaled up to megawatts.

This laser belongs to the family of high power chemical gas lasers. Principle of chemical lasers is that they efficiently convert energy derived from chemical bonds and reactions into excited states of atoms or molecules and then into the laser beam. These chemical reactions usually take place in gaseous media. Very huge advantage of the chemical lasers is then the high quality of the beam. This is due to the homogeneity of the lasing media in a resonator (usually low pressure gas). COIL has also the advantage that it can work either in continuous or in pulsed regime.

Apart from COIL there exists another type of the chemical gas laser, which is HF or DF laser, which operates at wavelength 2,7 μm. More about chemical lasers can be found for example in Perram.

In the COIL atomic iodine is pumped by near resonant energy transfer from excited oxygen molecule $O_2\left({}^1\Delta_g\right)$ (so called singlet delta oxygen) to atomic iodine in a ground state. Singlet oxygen itself is unsuitable for lasing as it has very long life time $(\sim 45\,\text{min})$. This is however the advantage for the transfer of singlet oxygen in the device.

Both the singlet oxygen and atomic iodine can be produced by different methods with different efficiency. Our task is to maximize the efficiency of both these processes and therefore improve the COIL operation.

Experimental Setup

COIL device consists of three main parts: a) singlet oxygen generator, b) supersonic nozzle, where chemical components are introduced and atomic iodine is produced and c) laser cavity with resonator, where the beam of light is produced by stimulated emission of radiation on the $I({}^2P_{1/2}) \rightarrow I({}^2P_{3/2})$ transition. Experimental setup can be seen in fig.

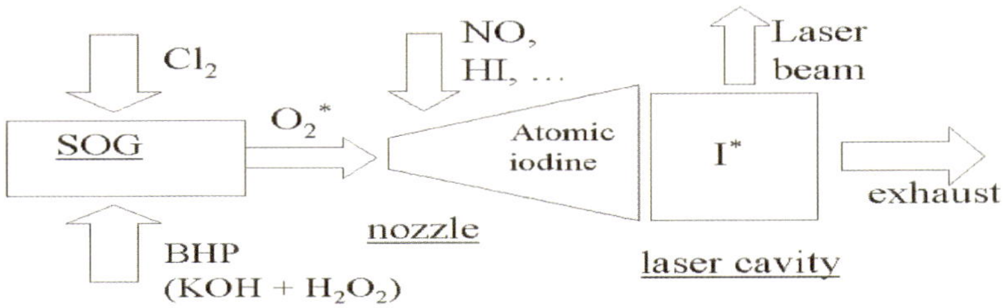

Figure: The experimental setup of the COIL laser.Singlet oxygen is produced in the singlet oxygen generator (SOG) in the reaction (2). Atomic iodine is then produced in the nozzle in the reaction (3) or (12). Then the energy is transferred through the reaction (23) from singlet oxygen to atomic iodine. In the laser cavity either laser beam is produced or diagnostics is applied to measure the concentration of I^*.

Singlet Oxygen Generation

In singlet oxygen generator oxygen in the $O_2\left({}^1\Delta_g\right)$ state is produced. The yield of singlet oxygen is defined as a ratio of concentrations of singlet oxygen and all oxygen present in generator:

$$Y = \frac{\left[O_2\left({}^1\Delta_g\right)\right]}{\left[O_2\right]}$$

We need to have high enough yield (over 15%) for the good laser operation, but the required quantity depends also on the method of I atoms generation.

There are many types of singlet oxygen generator (SOG). Classical method of generation is the chemical reaction between BHP (basic hydrogen peroxide - mixture of H_2O_2 and KOH) and chlorine Cl_2, which is usually diluted by N_2 or H_e:

$$Cl + 2KOH + H_2O_2 \rightarrow O_2\left({}^1\Delta_g\right) + 2KCl + 2H_2O$$

It is the liquid-gas surface reaction (BHP being liquid), so the possibly largest surface of liquid is needed for the contact of liquid and gas components.

Singlet oxygen can also be generated in electric discharge. Different types of discharge (RF, DC, micro wave) can be used in generator. One of the greatest problems with discharge SOG is the difficulty to sustain high enough yield of singlet oxygen at higher pressures due to the quenching of this state with atoms and molecules created in discharge (most harmful are N, O and O_3).

Atomic Iodine Generation

Originally, from the beginning of the COIL development, atomic iodine has been produced from iodine molecule I_2 which is dissociated by the singlet oxygen molecule. For one iodine molecule dissociation about 4–6 oxygen molecules are consumed. That diminishes the fraction of molecules able to pump the iodine and so diminishes the COIL efficiency. Another problem is that I_2 is really strong quencher of I^*. I_2 is also solid at the room temperature and must be heated to sublimate to gas phase.

Another method which has been developed is the discharge generation of atomic iodine from different molecules such as I_2, HI or CH_3I. Different types of discharge have been and still are examined, such as DC, RF or microwave discharge.

New method of atomic iodine generation was investigated. It is the chemical reaction of hydrogen iodide HI with atomic chlorine or fluorine.

Atomic Iodine Generation via Chlorine

Atomic iodine generation in the reaction of hydrogen iodide with atomic chlorine is described in the following equation:

$$HI + Cl \rightarrow HCl\left(HCl^*\right) + I, \qquad k = 1.64 \times 10^{-10}\, cm^3 molec^{-1} s^{-1}$$

Here k is the rate constant of this reaction and abbreviation molec stands for the molecule. First there must be atomic chlorine produced in the reaction with NO:

$$ClO_2 + 2NO \rightarrow Cl + 2NO_2$$

This actually is not simple chemical reaction, but the chain of more reactions and there are more reactions describing mainly Cl atoms losses (reactions:

$$ClO_2 + NO \rightarrow NO_2 + ClO, \qquad k = 3.4\times10^{-13}\, cm^3\, molec^{-1} s^{-1}$$
$$ClO + NO \rightarrow NO_2 + Cl \qquad k = 1.7\times10^{-11}\, cm^3 molec^{-1} s^{-1}$$
$$Cl + ClO_2 \rightarrow 2ClO, \qquad k = 5.9\times10^{-11}\, cm^3 molec^{-1} s^{-1}$$
$$Cl + NO_2 + He \rightarrow ClNO_2 + He, \qquad k = 7.2\times10^{-31}\, cm^6 molec^{-2} s^{-1}$$
$$ClO + NO_2 + He \rightarrow NO_3Cl + He, \qquad k = 1.0\times10^{-31}\, cm^6 molec^{-2} s^{-1}$$
$$Cl + Cl + He \rightarrow Cl_2 + He, \qquad k = 6.4\times10^{-33}\, cm^6 molec^{-2} s^{-1}$$
$$ClNO_2 + Cl \rightarrow Cl_2 + NO_2, \qquad k = 3.0\times10^{-11}\, cm^3 molec^{-1} s^{-1}$$

Losses of Cl atoms occur mostly in the termolecular reactions (reactions of three molecules). Importance of the termolecular reactions increases with pressure and decreases with temperature.

Atomic Iodine Generation via Fluorine

Like atomic chlorine also atomic fluorine can be used for I atoms generation. Here is the reaction of atomic fluorine with HI molecule:

$$HI + F \rightarrow HF\left(HF^*\right) + I, \qquad k = 1.51\times10^{-10} e^{-608.3/T} cm^3 molec^{-1} s^{-1}$$

Atomic fluorine is created from the molecular fluorine F_2 in reaction:

$$F_2 + NO \rightarrow F + NOF\left(NOF*\right), \quad k = 7\times10^{-13} e^{-1150/T} cm^3 molec^{-1} s^{-1}$$

This is not the only one process taking place during the fluorine atoms generation, there are some more reactions involving atomic fluorine, iodine, NO and diluting gas (helium in this case, fluorine must be diluted for safety reasons). Following reactions describe fluorine atoms losses:

$$F + NO + He \rightarrow NOF + He, \qquad k = 1.1\times10^{-31}\, cm^6\, molec^{-2} s^{-1}$$
$$F + F + He \rightarrow F_2 + He, \qquad k = 5.3\times10^{-34}\, cm^6 molec^{-2} s^{-1}$$
$$F + NO + NO \rightarrow NOF + NO, \qquad k = 1.7\times10^{-31}\, cm^6 molec^{-2} s^{-1}$$
$$F + I_2 \rightarrow IF + I, \qquad k = 4.3\times10^{-10}\, cm^3 molec^{-1} s^{-1}$$

And in following reaction iodine atom losses are described:

$$I + I + He \rightarrow I_2 + He, \qquad k = 3.8\times10^{-33}\, cm^6\, molec^{-2} s^{-1}$$
$$I + I + I_2 \rightarrow I_2 + I_2, \qquad k = 3.7\times10^{-30}\, cm^6 molec^{-2} s^{-1}$$
$$I + NO + He \rightarrow INO + He, \quad k = 5.5\times10^{-33}\, cm^6 molec^{-2} s^{-1}$$
$$I + INO \rightarrow I_2 + NO \qquad k = 2.6\times10^{-10}\, cm^3 molec^{-1} s^{-1}$$
$$I + F_2 \rightarrow IF + F, \qquad k = 1.9\times10^{-14}\, cm^3 molec^{-1} s^{-1}$$

There can be also included the reaction of atomic iodine with the wall, which causes another losses of iodine atoms.

Equations described here are then used for modeling of the chemical reactions (chemical kinetics). For the concentration estimation we use the fourth order Runge Kutta algorithm. As we now prepare mainly the experiments with atomic iodine generation via fluorine, this system is of great importance for us.

Applications

COIL technology has gained importance over the past few years owing to its nearly diffraction-limited optical quality, high scalability and short, fiber-deliverable wavelength. These distinct characteristics have made COIL a suitable technology for nuclear warhead dismantlement and nuclear reactor decommissioning. It is used as a weapon laser for the advanced tactile laser programs and military airborne laser. In addition, COIL can also be used for commercial or industrial applications which demands non-invasive, precise and rapid drilling or cutting.

Other applications of COIL include:

- Paint stripping;
- Rock crushing;
- Dismantling of nuclear facilities;
- Cleanup and survivor rescue during disaster;
- High power fiber optic transmission.

ElectricOIL

The Electric Oxygen Iodine Laser (EOIL) uses an electric discharge of a flowing oxygen gas mixture to generate singlet oxygen metastables, $O_2\left(a^1\Delta_g\right)$, and atomic oxygen, which subsequently react with molecular iodine to excite the atomic iodine lasing transition, $I\left({}^2P_{1/2} \rightarrow {}^2P_{3/2}\right)$ at 1.315 μm. The viability of EOIL has been recently demonstrated through measurements of positive gain and lasing in low-power laboratory systems.[1-4] The $I({}^2P_{1/2})$ (or I*) excitation mechanism in EOIL is similar to that for the Chemical Iodine Oxygen Laser, COIL, except that dissociation of the reagent I_2 occurs through rapid reactions with O rather than the much less efficient energy transfer from $O_2(a)$. COIL systems use an aqueous chemical process to generate $O_2(a)$, so no O is present, and I_2 must be dissociated by a complex multistep process which consumes some of the $O_2(a)$. EOIL's potential for gas-phase electric discharge generation of the active oxygen species offers substantial improvements in efficiency and weight limitations of closed-cycle systems.

The basic EOIL mechanism for the reactions between I_2 and the effluent of the oxygen discharge is:

$$O + I_2 \rightarrow IO + I \qquad I_2 \text{ dissociation}$$

$$O + IO \rightarrow O_2 + I$$

$$O_2(a) + I W O_2(X) + I^* \qquad I^* \text{ excitation (near-resonant, reversible)}$$

$$O + I^* \rightarrow O + I \qquad I^* \text{ quenching}$$

The two dissociation reactions have near gas-kinetic rate coefficients, and rapidly react in less than the reagent mixing time to produce complete dissociation if $[O] > 2[I_2]$. As with COIL, the I* excitation by energy transfer from $O_2(a)$ is near-resonant and reversible; the forward and reverse rate coefficients stand in the ratio of the temperature-dependent equilibrium constant:

The quenching of I* by O is not well understood, and may occur through a multistep reaction mechanism; however, the process results in a significant loss of I* (and thereby in $O_2(a)$) for typical discharge-generated O concentrations.5 Thus it is important to carefully control the O concentration through titrations with NO_2 and/or NO,[1-4] so that [O] is large enough to dissociate I_2 but small enough to minimize I* quenching:

$$O + NO + M \rightarrow NO_2 + M$$

$$O + NO_2 \rightarrow NO + O_2$$

(Additional reactions involving discharge-generated O_3 are possible, but are not likely to contribute owing to the low concentrations expected for O_3. Further investigations are in progress to test this contention.)

The I* excitation mechanism results in a rate law of the form,

$$d[I^*]/dt = k_f[O_2(a)][I] - k_r[I^*][O2(X)] - k_O[I^*][O]$$

Where, the net effect of I* quenching is to convert $O_2(a)$ to $O_2(X)$ as the reaction time increases. For slow quenching, a quasi-steady state approximation for [I*] gives,

$$[I^*]/[I] = k_f[O_2(a)]/\{k_r[O_2(X)] + k_O[O]\}$$

As $k_O[O]$ decreases through reduction of [O], $k_O[O] < k_r[O_2(X)]$, the expression approaches a true steady-state relationship,

$$[I^*]/[I] = k_f[O_2(a)]/k_r[O_2(X)] = K_{eq}(T)[O_2(a)]/[O_2(X)]$$

This expression defines the maximum [I*]/[I] ratio that can be achieved for a given $[O_2(a)]/[O_2(X)]$ and temperature. Through consideration of the atomic iodine state dynamics and degeneracies, it can be shown that population inversion and positive gain are achieved if [I*]/[I] > 0.5. Thus above eq. can be used to define a minimum $[O_2(a)]/[O_2(X)]$ ratio required to achieve gain.

The total O_2 introduced into the system, $[O_2]o$, is given by:

$$[O_2]_o = [O_2(X)] + [O_2(a)] + [O]/2$$

The ratio $[O_2(a)]/[O_2]o$ is then defined as the yield of $O_2(a)$ produced by the discharge, Y_Δ, and is a fundamental metric for the performance of the system. We can combine just above given two Equations to define the minimum $O_2(a)$ yield that must be exceeded in order to produce positive gain, in the limit of negligible [O]. The resulting curve is shown in figure. For a supersonically cooled flow of ~200 K, Y_Δ > 8.2% is required to achieve gain. However, in microwave discharge systems in the 0.1-1 kW range, we typically observe $O_2(a)$ yields in the 20-35% range for dilute O_2/rare-gas mixtures, well above threshold even at room temperature. Indeed, for a 70-100 W microwave discharge-flow system, we have observed positive $I* \rightarrow I$ gain in subsonic flow at 350 K[4].

The total $O_2(a)$ power produced by the EOIL discharge is the product of the $O_2(a)$ molar flow rate and energy:

$$P_\Delta = F_{O_2} \; Y_\Delta \; E_\Delta$$

Where, F O_2 is the molar flow rate of O_2, Y_Δ is the $O_2(a)$ yield, and EΔ is $O_2(a)$ energy, 94.369 kJ/mole. An alternative metric, which requires specification of the laser cavity temperature, is the $O_2(a)$ power available above the gain threshold:

ü§üüüüüüüüüüü$_{\Delta, avail}$ $_{O_2}$ $(\;_\Delta - \;_o(\;)) \;_\Delta = \;_\Delta (\;\; _o(\;) \;_\Delta)$

Where, $Y_o(T)$ is the threshold $O_2(a)$ yield as given by equations:

$[I^*]/[I] = k_f[O_2(a)]/k_r[O_2(X)] = K_{eq}(T)[O_2(a)]/[O_2(X)]$ and $[O_2]_o = [O_2(X)] + [O_2(a)] + [O]/2$ plotted in figure below. The available power is the theoretical maximum that can be extracted as laser power if there are no losses in the system. Typically, Yo(T) is greatly reduced by supersonic expansion, so the available power increases with decreasing cavity temperature. Limitations of reagent mixing, reaction kinetics, and optical losses will reduce the power extracted in a practical system. However, the total and available $O_2(a)$ power are important figures of merit in the evaluation of EOIL generator performance. Clearly, the performance scales as the product of the $O_2(a)$ yield and the total oxygen flow rate, so both high yields and high gas flow rates are required to scale laser powers into the kW range and higher. For example, if YΔ ~ 25% and

Yo(200 K) ~ 8%, then PΔ, avail ~ (2/3) PΔ, so an available laser power of 1 kW would require a total O_2(a) power of 1.5 kW and an oxygen flow rate (in rare gas diluent) of ~0.06 mole/s (80 l/min at STP).

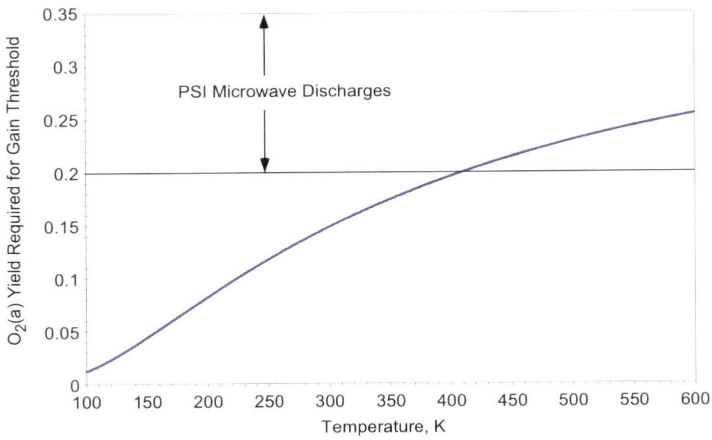

Figure: O_2(a) yields required to reach the I*/I transparency threshold

All Gas-phase Iodine Laser

In a chemical oxygen-iodine laser (COIL), the hydrogen peroxide solution acting as the energy source is rather bulky to carry in the mobile body. In addition, the solution cannot be stored for a long time as it tends to decompose. All this imposes certain restrictions to the application of COIL. Hence, all gas-phase iodine laser (AGIL) was developed as an alternative to mitigate the issues with chemical nature of COILs. It is a chemical laser in which gaseous iodine acts as a lasing medium.

The laser is operated based on the energy transfer between metastable chloronitrene (NCl) and ground state iodine molecules. During operation, chlorine atoms react with gaseous hydrazoic acid to produce excited NCl molecules that pass their energy to iodine atoms, similar to what singlet oxygen does in COIL. The excited iodine molecules further undergo standard laser emission.

AGIL has numerous advantages over other types of aqueous-based chemical oxygen iodine lasers, including zero-gravity operation, heat rejection in the exhaust and light weight characteristics.

Applications

AGIL is an ideal technology for aerospace applications as gases are lighter than liquids and convenient to maintain, store and transport.

HF (DF) Laser

The hydrogen fluoride (HF) laser is an infrared chemical laser that combines hydrogen produced in a combustion chamber with fluorine gas produced by thermal decomposition of compounds to create excited HF molecules.

It has the potential to provide up to several megawatts of continuous output. The laser outputs light at a wavelength of 2.7-2.9 μm. This wavelength can be absorbed by the air, thereby attenuating the beam and minimizing its range, and hence it has to be used in a vacuum environment.

The deuterium fluoride (DF) laser is a chemical laser formed by a mixture of fluorine and deuterium gas under controlled conditions. The wavelength of light produced by DF laser is longer than that of other conventional HF lasers, and hence it facilitates more effective laser transmission.

During its operation, ethylene is burned in the presence of nitrogen trifluoride in a combustion chamber, which produces free excited fluorine radicals. The mixture of deuterium and helium gas is then injected into an exhaust stream via a nozzle. The deuterium molecules react with the fluorine radicals to produce excited deuterium fluoride. The excited molecules further undergo stimulated emission in the optical resonating laser region.

A stable I-IF(DF) laser is also needed for use as the local oscillator in a laser-sun heterodyne radiometer detection array employed in high-resolution atmospherlc-transmis sion measurements.

HF (DF) Laser Design

The laser is shown in figure below. The device can be divided conveniently into the five regions (i. e., the discharge region, inlet region, lasing zone, heat exchanger, and exhaust). The discharge region provides an F-atom flow for the chemical reaction by means of a high-voltage discharge in a water-cooled pyrex tube containing SF 6 , Feo, and o2 that generates vibrationally excited HF or DF molecules. The He serves as a thermal diluent (and also reduces the required discharge voltage), whereas the SF 6 is the F-atom source. Oxygen is added to the flow toincrease the Fatom concentration and to eliminate S deposition through exothermic reaction with the S atonis formed in the discharge. The mixed gases are injected into one end of the discharge tube upstream of a group of eight symmetrically deployed t/t6-in. -diameter by 6-in.-long Ni wires that serve as discharge anodes. The operation of the Ni wires as cathodes resulted in metal loss and reduced power output. The discharge uniformity depends heavily upon the flow uniformity of the gases at the electrode tips; therefore, the gas injection into the discharge tube is made as uniform as possible. The long anodes are gas-cooled in the region upstream of the discharge. The Ni anodes are individually connected through ballast resistors (50 kO, 100 ko, or 200 kD) to the positive terminal of the power supply. The ballast resistor values can be varied to accommodate power-supply characteristics and laser operating conditions. The ballast resistor bank is forced-air-cooled.

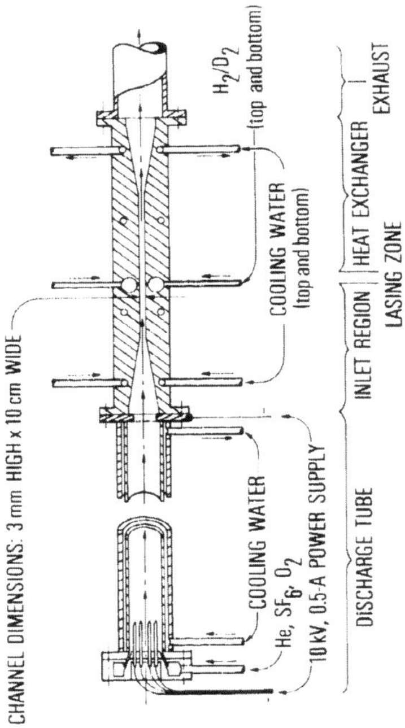

The discharge tahes place along a central core in the pyrex tube and terminates on a water-cooled Cu hole electrode (3/4-in. diameter) at the entrance to the laser inlet region. The pyrex tube has a l-in. i. d., o. 062-In. wall thickness and nominally is used in two lengths, 12 and 24 in., with 10- and Zo-kV power supplies, respectively. An Al$_2$ o 3 cylindrical shell is inserted inside the pyrex tube at the junction with the cathode to eliminate discharge attachment at the pyrex-Cu cathode interface. The water jacket surrounding the pyrex tube is made of lucite.

The hot flow containing r atoms passes through the cathode hole into the inlet region where the flow cross section is changed from a square gnomatry of the manie nominal cross section as the discharge tube to the channel crosN section 10 by 0.3 cm. The H_2 is injected along a line transverse to the flow at both the top and bottom of the channel through 40 evenly spaced 0.0135-in.-diametorholesin each wall. The H_2 is fed from plenums Immediatoly above and below the injection hole array. The lasing zone is immediately downstream of the It2 injection ports, where the mixing of the H_2 and F takes place. The remaining portion of the channel downstream of this region and the transition section in which the rectangular channel geometry is returned to a circular geometry for convenience in ducting serve as a heat exchanger to reduce the temperature of the heated gas to -200°C. Further temperature reduction of the gas is accomplished in a downstream heat exchanger.

The inlet region, the mixing arid lasing zone, and the heat exchanger are fauricated in a single Amsulf Cu block 8 by 4 by I -1/2 in. The assembly is water-cooled and contains the H_2 plenums. The laser is exhausted to a 150 cfni vacuum pump through a conventional 1-1/2 in. i.d. flexible vacuum hose. A surge tank in the exhaust is unnecessary for damping pump pressure fluctuations since no pressure variations at the laser are detectable with the un damped system operating at normal flow rates. Brewster-angle BaF_2 windows are mounted on channel sidewalls that are o-ring sealed to the channel body. The windows are centered along the gain region in the channel, nominally, I-- mm downstream of the 14 injection holes. Laser-beam polarization can be controlled by the rotation of the Brewster-angle windows for proper orientation with respect to the gratings. In an early experiment, a small purge flow of He was injected at the Brewster windows to reduce the accumulation of ground-state HF or DF, but the use of the flow did not affect laser performance and, hence, was discontinued. The optical cavity is constructed from Burleigh optical erector set components and consists of stainless-steel star gimbal mounts on stainless-steel rod plates that arc clamped to four I-in. -diameter Invar rods. The laser Cu body is suspended between the optical cavity rods with the gain region co linear with the optical cavity centerline. Multiline power can be coupled from the laser by means of a partially transmitting mirror used in conjunction with a totally reflecting mirror, one of which is a spherical mirror of radius I to 10 m and the other, an optical flat. Single-line operation can be accomplished by use of a 300- or 600-lines/mm grating in the auto collimation position with zero-order coupling for lasing at the desired wavelength and with a spherical totally reflecting mirror used as the other cavity component.

MIRACL

The Mid-Infrared Advanced Chemical Laser (MIRACL) is the first megawatt-class, continuous wave, chemical laser built in the free world. It is a deuterium fluoride (DF) chemical laser with energy spectra distributed among about ten lasing lines between 3.6 and 4.0 microns wavelength. Since it first lased in 1980, it has accumulated well over 3500 seconds of lasing time. To date, it is the only laser produced in the US, at DF wavelength, which has attained average powers at this high level.

MIRACL operation is similar to a rocket engine in which a fuel (ethylene, C_2H_4) is burned with an oxidizer (nitrogen tri fluoride, NF_3). Free, excited fluorine atoms are one of the combustion products. Just downstream from the combustor, deuterium and helium are injected into the exhaust. Deuterium reacts with the fluorine to yield excited deuterium fluoride (DF) molecules, while the helium stabilizes the reaction and controls the temperature. The laser's resonator mirrors are mounted outside the excited reaction gas region and optical energy is extracted. The cavity is actively cooled and can be run until the consumable supply is exhausted.

The laser's output power can be varied from full power down to 13% by altering the reactant flow rates. Resonator cavity pressure at full power is less than 100 torr. Exhaust gases flow through the laser resonator cavity to either one of two rectangular diffusers, where the pressure is passively recovered to 200 torr. This is followed by scrubbing, where the hot effluent is washed by a profuse shower of fine water droplets. The remaining pressure is recovered to atmospheric pressure in two 100 meter-long steam ejectors and the remaining cleaned gases and steam from the ejectors are dispersed to the atmosphere.

Lasing occurs between the cavity's concave and convex resonator optics arranged in a confocal unstable configuration. Out coupling of the optical energy is accomplished by a scraper mirror which is set at an angle of 45 degrees to the optical axis of the cavity. Next, the laser beam is reflected through a three-element beam-shaping telescope and an aerodynamic window to pass the beam from the cavity pressure to atmospheric pressure. The optical path between the cavity and the beam director has 11 high power mirrors. The path is conditioned with Argon or Nitrogen gas to prevent thermal beam distortion.

A low efficiency line grating on a relay optic is used to provide a small sample of the MIRACL's beam to a suite of beam diagnostic instrumentation, and to angle measurement sensors for automatic alignment of the beam. The angle measurement is made by two quad detectors; one for the MIRACL beam, the other for a helium-neon pilot beam. The beam-angle detectors drive steering mirrors to maintain proper alignment of the resonator and the optical train mirrors. The helium-neon detector also senses the alignment of a separate tower alignment beam from the beam director and drives a steering mirror at the base of the SLBD's tower. This Tower Beam Alignment System (TBAS) maintains the high energy laser beam alignment to the Sea Lite Beam Director. In this way, the entire optical train is maintained in alignment from end to end.

3

Soild State Lasers

A solid-state laser is any laser that uses a solid gain medium. The diverse types of solid state lasers such as diode-pumped solid-state laser, thin-disk laser, Er:YAG laser, ruby laser, fiber laser, dye laser, etc. have been thoroughly discussed in this chapter.

Solid-state lasers are lasers whose active medium is typically a minority ion in a solid-state host. The host is most often a single crystal with about 1% of a different species, such as a neodymium ion (Nd^{3+}), doped into the solid matrix of the host. For certain applications the host is a glass, the most important of these being glass doped with the erbium ion (Er^{3+}), which is then pulled into a long thin fiber and used in telecommunications systems. Solid-state lasers use optical pumping by other lasers or lamps to produce a wide variety of laser devices. The population inversion (a condition in which the atomic population of the higher energy state is greater than the lower energy state) in solid-state lasers is created when the active medium absorbs photons from an intense light source.

One important class of solid-state lasers consists of the tunable and ultrafast lasers. These systems use solid-state lasers to excite them, and have other similarities in their thermal and optical properties.

The typical optically pumped, solid-state laser system is much larger than a semiconductor device. The optical pump may either be a single, "high-power" semiconductor diode, a small array of semiconductor diodes on one chip, or a stack of arrays for a laser with outputs of up to 1000 W. Before the advent of laser diodes, solid-state lasers were pumped by lamps—either running continuously for modest power lasers, or pulsed flash lamps for laser output energy in the 1-Joule-per-pulse category. Since the number of arrays needed to power a 1-J-per-pulse laser is still quite pricey, lamps still have an important niche in the solid-state laser business.

Characteristics of Solid State Lasers

Continuous Wave Type

The term continuous wave or CW normally indicates that a laser has continuous output. Usually in CW case of laser is continuously pumped.

There are very few solid state crystals that can produce laser light and resist the extreme heat generated by a CW pumping source. Only common laser of this type is the Nd: YAG. Output wave of the Nd: YAG laser is 1.06 microns which is in the near infrared spectrum.

The beam diameter varies from 0.75mm to 6mm while beam profile can either TEM0, 0 or multimode. The TEM0,0 beams tend to have narrower beam diameters.

Importantly not that the beam divergence from a solid state laser is not constant. As the laser rod gets heated by the light source during pumping, thermal expansion occurs. This causes the rod to act as a lens and expand beam. So during use when the rod gets hotter than the rod expands even more and so causing greater divergence.

The beam divergence for the CW solid state laser can be as low as 1 milli radian for TEM0,0 mode lasers are as high as 20 milli radians for multimode lasers.

The output power for the CW solid state laser varies from a low of 0.4 watt to a high of 600watts.

Since a Q-switch can be inserted between the laser mirrors, therefore a very rapidly pulsed output can be obtained. In addition the CW pumped solid state lasers can be mode locked for the production of ultra short pulses.

In fact these lasers can produce a CW, modulated CW, Q-switched or mode locked output.

Pulsed Type

The pulsed solid state laser produces a pulsed output due to the pulsing of the input energy. In pulsed lasers the material used can be cooled between pulses therefore the active medium does not exposed to the extreme temperature rise, experienced by a CW laser. That is why there are more choices of active media for pulsed laser rather than for CW solid state lasers.

When Nd: YAG is used as a active medium then it can pulsed very rapidly because it can even with stand the extreme thermal loads of CW operation. It is very common to pulse YAG lasers a rate of up to 200 pulses per second. However if ruby or Nd: glass is used as the active medium then pulse rates are limited to 2 or 3 pulses per second. The additional time between pulses is needed to cool the crystal to prevent fracturing.

The Nd: YAG and ND: glass put out the wavelengths in the near infrared spectrum, i.e. 1604 and 1600nm respectively. Ruby laser emits a visible beam at 694nm (lightly red).

Beam diameter of pulsed laser is in 5 to 10mm range. Beam divergence ranges from 1 milli radian to 10 milli radian. Power output from pulsed laser averages about 400watts although the peak power of individual pulse is much higher.

Construction of Solid State Lasers

The typical solid state laser construction can be shown as:

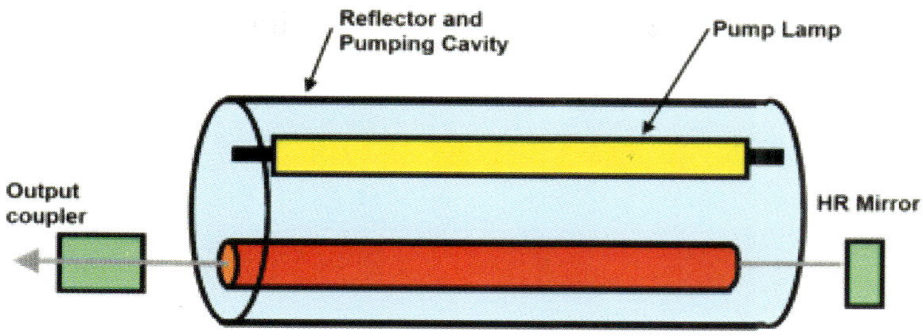

Usually all solid state lasers have similar design. A laser rod is mounted near an arc or flash lamp. The lamp is connected to DC power supply that maintains a controlled current through lamp. The laser rod and lamp are placed parallel to each other and are surrounded by a reflector. The high reflective (HR) mirror and the output coupler are placed at either ends of the laser cavity.

For simplicity we do not show the DC power supply and cooling system, however they exists in solid state lasers.

Function of Solid State Lasers

We know that active medium used for solid state lasers is a solid material. Usually all solid state lasers are pumped optically means light source is used as energy source of solid state lasers. When the solid material rod absorbs the light energy from light source then becomes excited. The upper energy level of the ions are radiation less, however when the energy transition takes place then the metastable upper laser level is soon reached. At this point the emissions occurs which obviously results in lasing.

By energy level diagram the function of solid state say Nd: YAG laser can be shown as:

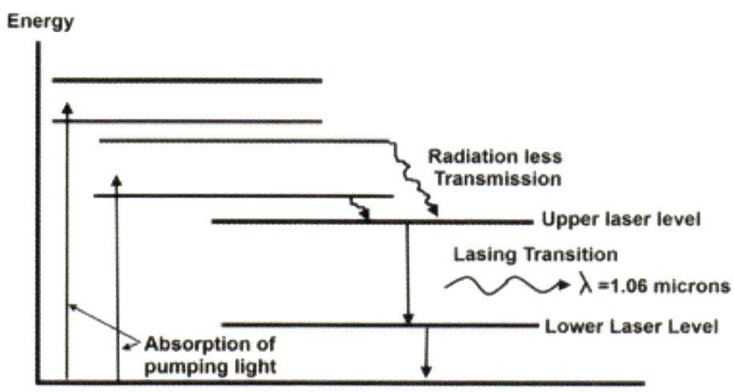

Wavelength Tuning

In terms of their potential for wavelength tuning, different types of solid-state lasers differ considerably. Most rare-earth-doped laser crystals, such as Nd:YAG and Nd:YVO_4, have a fairly small gain bandwidth of the order of 1 nm or less, so that tuning is possible only within a rather limited range. On the other hand, tuning ranges of tens of nanometers and more are possible with rare-earth-doped glasses, and particularly with transition-metal-doped crystals such as Ti:sapphire, Cr:LiSAF and Cr:ZnSe (→ *vibronic lasers*).

Types of Solid-State Lasers

Examples of different types of solid-state lasers are:

- Small diode-pumped Nd:YAG (→ *YAG lasers*) or Nd:YVO_4 lasers (→ *vanadate lasers*) often operate with output powers between a few milliwatts (for miniature setups) and a few watts. Q-switched versions generate pulses with durations of a few nanoseconds, micro joule pulse energies and peak powers of many kilowatts. Intracavity frequency doubling can be used for green output.
- Single-frequency operation, typically achieved with unidirectional ring lasers (e.g. NPROs = nonplanar ring oscillators) or microchip lasers, allows for operation with very small linewidth in the lower kilohertz region.
- Larger lasers in side-pumped or end-pumped configurations having the geometry of rod lasers, slab lasers or thin-disk lasers, are suitable for output powers up to several kilowatts. Particularly thin-disk lasers can still offer very high beam quality, and also a high power efficiency.
- Q-switched Nd:YAG lasers are still widely used in lamp-pumped versions. Pulsed pumping allows for high pulse energies, whereas the average output powers are often moderate (e.g. a few watts). The cost of such lamp-pumped lasers is lower than for diode-pumped versions with similar output powers.
- Fiber lasers are a special kind of solid-state lasers, with a high potential for high average output power, high power efficiency, high beam quality, and broad wavelength tunability.

Applications/Uses of Solid State Lasers

- Nd: YAG solid state lasers usually used when drilling holes in metals.
- Nd: YAG pulsed type solid state lasers can be used in medical applications such as in endoscopy etc. As military application.
- Nd: YAG is used by target destination system.

Advantages of Solid State lasers

- The solid state lasers have the following advantages over other types of lasers.
- No chance is of wasting material in the active medium because here material used is in solid form not in gas form, where this occurs.
- Both continuous and pulsed output is possible from solid state lasers.

- Solid state lasers have high efficiency from some of gas lasers such as He-Ne lasers and Argon Lasers efficiency of solid state Nd: YAG laser is 2% to 3%.
- Construction of solid state laser is comparatively simple.
- Beam diameter of solid state laser is very less than CO_2 lasers.
- Output power ranging from very low value of about 0.04 watts to high value of about 600 watts.
- Cast of solid state lasers is economical.

Disadvantages of Solid State Lasers

- Efficiency of solid state laser is very low as compared to CO_2 lasers.
- Great disadvantage of solid state lasers is the divergence, which is not constant and ranges 1 milli radian to 20 milli radian.
- Output power is also not very high as in CO_2 lasers.
- Due to thermal lasing in solid state lasers, the power loss occurs when the rod gets too hot.

Diode-pumped Solid-state Laser

Diode pumped solid state (DPSS) lasers are solid state lasers made by pumping a solid gain medium, for example, a ruby or a neodymium doped yttrium aluminum garnet (YAG) crystal, with a laser diode. DPSS lasers have advantages in compactness and efficiency over other types, and high power DPSS lasers have replaced ion lasers and flash lamp pumped lasers in many scientific applications, and are now appearing commonly in green and other color laser pointers. Fibre optic waveguides have also been doped with various rare earth ions and pumped by semiconductor lasers. Fibre lasers were realized shortly after the discovery of the laser and were initially transversely pumped by the crude but effective technique of coiling a fibre round a flash lamp. The resulting diode pumped laser has, in general, greatly improved spatial and spectral characteristics than the laser diode itself.

DPSS and diode lasers are two of the most common types of solid state lasers. However, both types have their advantages and disadvantages. Interest has increased in the past few years in using semiconductor diode lasers to excite solid state lasers based on rare earth ion doped transparent solids such as neodymium doped yttrium aluminum garnet (Nd: YAG). Figure below shows the absorption spectrum is for 1% doped Nd: YAG. The pulsed flash lamp emits radiation at all wavelengths while the diode laser emits radiation at essentially a single wavelength that can be tuned to a particular absorption line of the Nd: YAG. Traditionally, these solid state lasers are excited by flash lamps that emit broadband radiation. Lamp pumped systems are inefficient, however, with typically 1% electrical to optical efficiency, and the lamps need replacement after approximately 200 hours when operated continuously. Diode laser pump sources allow operation at higher efficiency (10%) and longer life (20.000 hours). DPSS lasers have, over the last ten years, become very important in the laboratory and are beginning to make an impact in the market place. During the last decade the increasing demand for miniaturisation of micro system devices and their associated components has established laser micromachining as the preferred fabrication method in many industrial sectors.

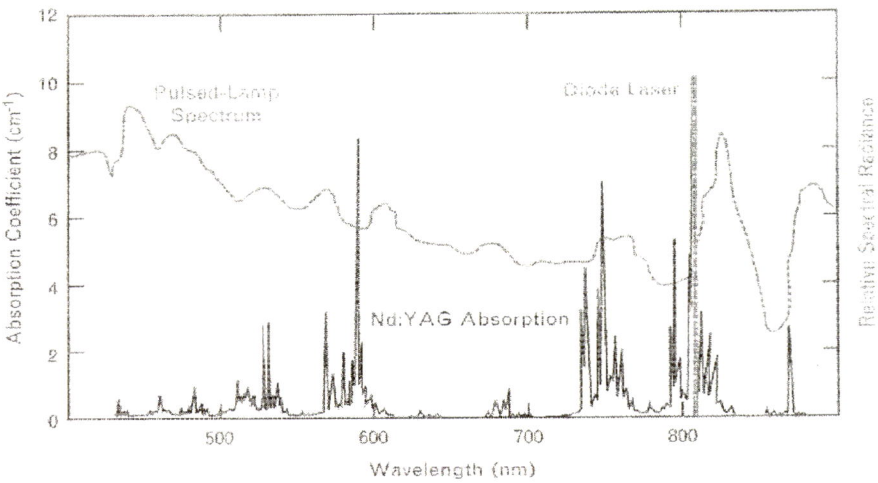

Figure: Absorption spectrum of Nd: YAG and the emission spectra of a diode laser and a pulsed flash lamp.

Characteristics of DPSS Lasers

- Optical efficiency: DPSS lasers are highly efficient because of the direct excitation of the pump beam into the useful absorption band of the lasing ion. Direct excitation minimizes the unwanted losses in the lasing crystal with optical to optical efficiency of up to 70%. Selecting the composition of the host material enables laser diodes to be constructed with wavelengths between 600 nm and approximately 30 μm in the infrared. Over much of this region the output power and lifetime of laser diodes is remarkably poor. Only in wavelength regions where strong commercial or military interest is present have sufficient resources been deployed to develop an efficient and long lived device. Notable examples of this process have occurred at 1.5 μm and 1.3 μm, which are important telecommunications wavelengths, and near 800 nm which is important for communications, entertainment and medical applications. Only in the vicinity of 800 nm have high powered devices been built. The driving force has been military applications including pumping Nd doped solid state lasers.

- Wavelength: The wavelength at which laser diodes operate is dictated by the size of the band gap, since the light arises from the recombination of electrons and holes in a pn junction. The band gap may be tuned in size by two main processes:

  1. Altering the composition of the host material.
  2. Changing the temperature of the host material. Other physical effects such as the application of pressure may also change the band gap but they tend to produce too small an effect to be useful. The output wavelength of diode lasers varies from diode to diode because of small differences in fabrication and the wavelength changes with temperature. The variation in output wavelength leads to increased cost because only diode lasers in a small wavelength range are usable. The change in wavelength resulting from temperature variation requires that the diodes must be temperature controlled.

- Operational lifetime: The operational lifetime of laser diodes or arrays is much larger than that of conventional arc or filament lamps. A typical laser diode array can operate without significant degradation for more than 10.000 hours, but usually up to $3x10^4$ hours, while a cw lamp must be replaced after 200-400 hours of operation (or 10^7 shots in the case of pulsed pumping). The performance of a diode laser degrades exponentially with time. Initially, the failure rate is low, but it increases exponentially with the operating time. Failure mechanisms of laser diodes are divided into two main classes:

  1. User induced damage (such as mechanical, thermal or electrical shock or electrostatic discharge).
  2. Intrinsic damage, which results from three main sources:

     - Degradation of laser mirrors or facets because of high current densities or current spikes. This will increase internal losses and lead to catastrophic failure.

     - Damage resulting from crystal defects within the active region, which will lead to an increase in absorption losses within laser diode. Such defects are common in AlGaAs laser diodes because of the oxidation and migration of the aluminum. At present efficient aluminum free laser diodes replace the aluminum containing lasers.

     - Resistive losses and heating because of increases in the resistance of electrical contacts to the laser.

- Temperature: Since the diode laser is narrow bandwidth pumping source, it pumps only the useful absorption bands relevant to laser action, reducing the thermal load in the crystal. This thermal load results from the quantum gap between the pump and the leasing photons. Thermal effects such as thermal lensing, thermally induced birefringence, and thermal damage to the lasing to the crystal are reduced significantly. Fine tuning the temperature of the laser diode causes a change in wavelength. For GaAlAs devices the wavelength tunes at an approximate rate of +0.25 nm^{o}C^{-1} mainly due to the change in band gap with temperature. This feature is used to tune the laser diode into coincidence with the absorption bands of rare earth ions. Normally the lasers will be cooled down, since this gives a longer lifetime for the laser diode. It is usual to specify the room temperature wavelength of the laser diode some 5 nm longer than the rare earth absorption feature that will be pumped. Simple Peltier coolers can provide a temperature change of about 40°C corresponding to a wavelength shift of 10 nm. The Nd: YAG absorption line width is ~2 nm so the diode temperature has to be set to better than ±4°C, which is not too demanding in the laboratory. The influence of temperature on the wavelength of the laser diode causes problems of packaging the DPSS laser. Pulsed laser diodes suffer from a transient thermal wavelength shift. Since the current is pulsed through the laser diode, the temperature is never in equilibrium and a transient wavelength shift occurs. The wavelength increases during the optical pulse. Measurements of this effect have revealed shifts of ~5 nm. Shifts of this magnitude are larger than the absorption line width of many solid state laser materials. Account has to be taken of this effect when predicting the efficiency of pulsed DPSS lasers. The diode lasers are normally cooled by thermoelectric cooler for low power systems, and by liquid cooling for high powers. Cooling can stabilize the diode laser frequency.

- Beam quality: Although the beam quality of a laser diode or diode array is not good, the use of coupling optics makes it possible to obtain a good TEM00 beam mode form a DPSS laser. The coupling optics circularize the output beam emanating from the laser diode array or bar, and then couple the beam into the solid state laser crystal either by direct coupling or an optical fiber. The cylindrical fast axis collimating lens can reduce the beam divergence of diode laser stacks to a value of lens than 10 mrad. The solid state laser can be pumped longitudinally or transversely. This subject is technologically well established and will be discussed in the next section. The laser crystal host can be in the form of a crystal, waveguide or optical fiber.

Since the absorption length of the diode laser beam focused inside the solid state laser crystal is short, the pump mode volume is smaller than the laser cavity mode volume and one expects a good spatial beam quality. The laser cavity itself is short and therefore the output power of the DPSS laser is a single longitudinal mode. The best transverse mode quality is obtained from single stripe devices. The low ellipticity of a source with dimensions of 3x1 μm ensures that the output power may be efficiently coupled into optical fibre or into solid state laser materials. Single laser diodes may provide up to 150 mW in a single transverse mode. The shape of the output beam from a single strip device is shown in figure below.

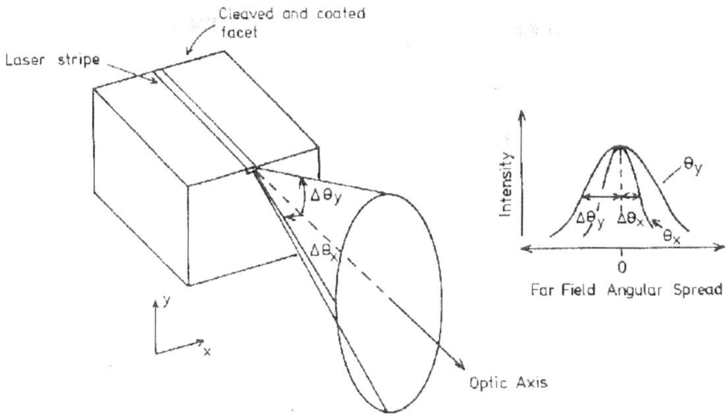

Figure: The elliptic beam shape arising from the transverse dimensions of a single laser diode stripe

To date the most extensively used pumping geometry is longitudinal pumping. The output from the diode laser is collimated and beam shaped to achieve a circular profile before being focused down to form a pump spot on the laser rod. This technique allows good matching between the lasing spot size of the solid state laser cavity and the pump spot size in the gain medium. This spatial overlap between pump and lasing modes, known as mode matching, is critical to the efficiency of the diode pumping process.

Frequently, DPSS lasers are pumped by laser diode arrays in order to achieve high output powers. As has been noted, the simplest way to increase the output power from a single laser diode is to increase the width of the emitting region either by fabricating a number of laser diode stripes in close proximity so that there is a series of emitting regions, or by enlarging the width of the electrically pumped region. The output of arrays lies in the region of 200 mW to 3 W cw output at present. True arrays of laser diode stripes tend to be partially coherent and may exhibit the two lobe structure in the far field characteristic of the phase changes due to evanescent coupling between adjacent stripes. This feature appears less obvious as the output power and number of stripes increase due to reduced coherence across the array. The broad stripe arrays are multi transverse mode devices and the beam quality depends on how hard they are driven. Applications of high power fiber delivery from diode lasers include pumping either bulk or fiber lasers, machining and marking, soldering, and power transmitting. High power fiber coupled diode laser sources could lead to the creation of compact high power diode pumped lasers. Similar systems that have been demonstrated either have relatively low power because they were limited to only one or two sources focused into a fiber, or because they consist of a fiber bundle in which one or two sources were focused into a fiber and then many fibers were brought together to form the pump source. A single fiber to carry all the power, as opposed to a fiber bundle, is desirable because it is less bulky and it maintains a greater degree of pump beam brightness, which allows improved performance from diode pumped lasers.

Common DPSS Processes

The most common DPSS laser in use is the 532 nm wavelength green laser pointer. A powerful (>200 mW) 808 nm wavelength infrared GaAlAs laser diode pumps a neodymium doped Nd:YAG or a neodymium doped yttrium orthovanadate (Nd: YVO_4) crystal which produces 1064 nm wavelength light from the main spectral transition of neodymium ion. This light is then frequency doubled using a nonlinear optical process in a potassium titanyl phosphate ($KTiOPO_4$, KTP) crystal, producing 532 nm light. Green DPSS lasers are usually around 20% efficient, although some lasers can reach up to 35% efficiency. In other words, a green DPSS laser using a 2.5 W pump diode would be expected to output around 500-900 mW of 532 nm light. In optimal conditions, Nd:YVO_4 has a conversion efficiency of 60%, while KTP has a conversion efficiency of 80%. In other words, a green DPSS laser can theoretically have an overall efficiency of 48%. In the realm of very high output powers, the KTP crystal becomes susceptible to optical damage. Thus, high power DPSS lasers generally have a larger beam diameter, as the 1064 nm laser is expended before it reaches the KTP crystal, reducing the irradiance from the infrared light. In order to maintain a lower beam diameter, a crystal with a higher damage threshold, such as lithium triborate (LBO), is used instead. Much of the excitement in nonlinear optics has been caused by a new generation of nonlinear materials. These new materials are non-hygroscopic, have high damage threshold and good phase matching characteristics. Materials which are now widely available include KTP, LBO potassium niobate (KNB) and beta barium borate (BBO).

- Blue DPSS lasers use a nearly identical process, except that the 808 nm light is being converted by an Nd: YAG crystal to 946 nm light (selecting this non principal spectral line of neodymium in the same Nd doped crystals), which is then frequency doubled to 473 nm by a BBO or LBO crystal. Because of the lower gain for the materials, blue lasers are relatively weak, and are only around 3-5% efficient. In the late 2000s, it was discovered that bismuth triborate (BiBO) crystals were more efficient than BBO and LBO and do not have the disadvantage of being hygroscopic, which degrades the crystal if it is exposed to moisture.

- Violet DPSS lasers at 404 nm have been produced which directly double the output of a 1.000 mW 808 nm GaAlAs pump diode, for a violet light output of 120 mW (12% efficiency). These lasers out perform 50 mW gallium nitride (GaN) direct 405 nm Blu ray diode lasers, but the frequency doubled violet lasers also have a considerable infrared component in the beam, resulting from the pump diode.

- Yellow DPSS lasers use an even more complicated process: A 808 nm pump diode is used to generate 1,064 nm and 1,342 nm light, which is summed to become 593.5 nm. Due to their complexity, most yellow DPSS lasers are only around 1% efficient, and usually more expensive per unit of power. In addition to optical fibre waveguides, waveguide lasers have been fabricated in bulk glasses and in a variety of crystals.

Examples of DPSS Lasers

In this section we consider the performance of a range of specific DPSS lasers. We begin by considering in some detail the laser diode pumped Nd: YAG laser, which is certainly the best studied laser diode pumped system. We will then consider the performance of the laser diode pumped Nd:X laser, where X is any one of a number of host materials and also laser diode pumped stoichimetric materials. The stimulus for the research into these materials is essentially to reduce the size of DPSS lasers.

Nd Doped DPSS Lasers

The trivalent neodymium ion was the first of the rare earth ions to be used in a laser and it is the dopant which has received most attention in the field of DPSS lasers. The Nd^{3+} ion has a strong absorption at 0.81 μm, which coincides with the emission wavelength of commercially available GaAs and GaAlAs laser diodes. Most of the research reported on laser diode pumped Nd doped lasers has concentrated on the four level high gain $^4F3/2$ - $^4I11/2$ transition which corresponds to a laser output in the region of 1.06 μm. However, work has also been reported on the 1.3 μm $^4F3/2$ - $^4I13/2$ transition, and on the three level 0.946 μm $^4F3/2$ - $^4I9/2$ transition. In this paper, all Nd doped laser results refer to the 1.06 μm transition unless otherwise stated shows a simplified energy level diagram for Nd and several other important rare earth ions.

The upper state lifetime is another important parameter for energy storage which can vary significantly with the choice of host medium. The problem which effects Nd^{3+} doped laser materials is that of concentration quenching of the upper state lifetime, where the lifetime decreases with increasing Nd^{3+} concentration. This leads to an increase in pump threshold. However, high doping is often desirable in DPSS lasers so that the pump radiation can be absorbed in a smaller volume, yielding a lower pump power threshold. One class of material in which this problem may be over come is that of stoichiometric materials, where the Nd^{3+} is a constituent component of the material, rather than a dopant.

Laser Diode Pumped Nd: YAG Laser

The Nd:YAG laser has become the most common DPSS laser for a variety of reasons. The advantages of the Nd^{3+} ion as a dopant have been mentioned previously. The level of neodymium doping in YAG is limited to about 1.5% due to concentration quenching of the upper state lifetime. The YAG host is hardness, high thermal conductivity and good optical quality.

The highest overall efficiencies are generally obtained from the longitudinally or end pumped geometry, due to the excellent matching that can be obtained between the solid state laser TEM00 mode. The first highly efficient laser diode end pumped Nd: YAG laser was reported by Sipes, who obtained an output of 80 mW at an overall (electrical to optical) efficiency of 8%. The limitations imposed on efficient coupling using end pumping mean that to obtain higher powers it is necessary to use a scheme whereby a laser rod or slab is pumped using a transverse pumping geometry. A disadvantage of the transverse pumped geometry tend to be much lower than for the end pumped case due to poor mode matching. Several authors have published results for pulsed systems. Optical slope efficiencies of 23% for slabs and 54% for rods have been obtained.

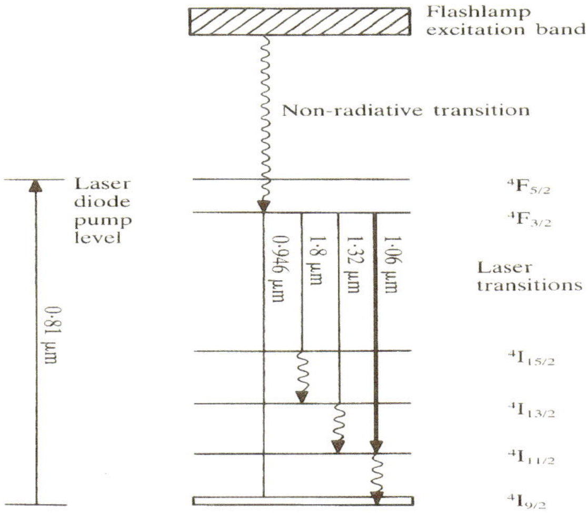

Figure: Enegy level diagram for Nd:YAG, illustrating process of diode laser excitation and Nd:YAG laser transitions

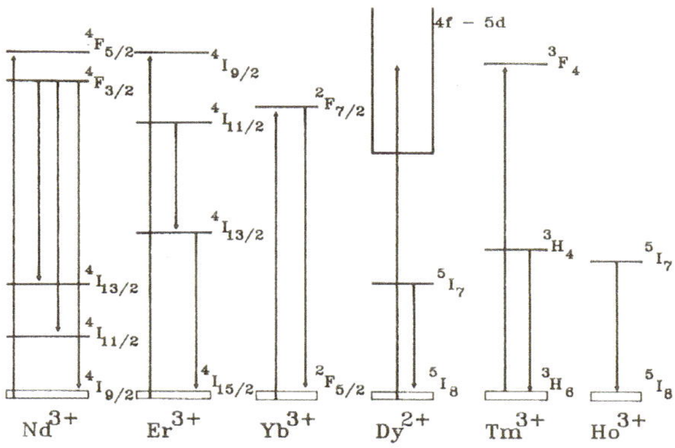

Figure: Simplified energy level diagrams for some of the most important rare earth ions

Transversely pumped cw Nd: YAG lasers have also been reported. Burnham and Hays used four cw laser diodes to pump a Nd:YAG rod transversely. 3.3 W of multimode output were obtained, at an overall efficiency of 3.5%. As expected, the efficiency of TEM00 operation was lower at 2%.

Laser Diode Pumped Nd:YLF Laser

The Nd: YLF laser has several properties very different from those of the Nd:YAG laser. The fact that it is a uniaxial material means that, simply by using an intracavity polarizer, one can select one of two different wavelengths for each transition. For the $^4F3/2$ – $^4I11/2$ transition these wavelengths are 1.047 μm, and 1.053 μm, with the 1.047 μm polarization exhibiting the higher gain. The natural birefringence of Nd: YLF swamps the effect of thermally induced birefringence observed in materials such as Nd: YAG.

The fluorescence lifetime of Nd: YLF is approximately twice as long as that of Nd: YAG. Furthermore, a large amount of energy can be stored in the medium gain 1.053 μm transition before the onset of amplified spontaneous emission. These facts make the laser diode pumped Nd: YLF laser an attractive medium for the generation of high power Q switched pulses. The results show that approximately twice the pulse energy can be obtained from a Q switched Nd: YLF laser than from a Q switched Nd: YAG laser, due to the difference in the fluorescence lifetime for these media. The highest peak power that has been obtained to date from the laser diode pumped Nd: YLF laser is 70 kW, in a pulse of duration <10 ns.

Q switching was demonstrated in light emitting diodes pumped Nd: YAG and Nd doped potassium gadolinium tungstenate (Nd: PDT). A number of different techniques for Q switching were used including electrooptic, acoustooptic, saturable absorber, and cavity dumping. Pulse lengths as short as 4 ns using cavity dumping were obtained, with the highest peak power being 170 W in a 65 ns pulse in Nd: PGT by acoustooptic mode locking.

Laser Diode Pumped Nd: Glass Laser

Nd: glass is an ideal candidate for laser diode pumping due to its wide absorption spectrum in the region of 800 nm. Unlike, for example, the YAG host, the concentration of active ions can be very high before the onset of concentration quenching (approximately 7% Nd_2O_3 in phosphate glasses). The main disadvantage of glass as a host medium is its low thermal conductivity (0.6 W $m^{-1}K^{-1}$ for Schott LG760 phosphate glass as opposed to 11 W $m^{-1}K^{-1}$ for Nd: YAG). This makes the Nd:glass laser particularly susceptible to thermal effects, such as thermal lensing, thermally induced birefringence and thermal damage. The first reported operation of an laser diode pumped Nd:glass laser was by Kozlovsky et al.. A pump power threshold of 2 mW and a slope efficiency of 42% was obtained from a monolithic device, when pumped with a low power single stripe laser diode.

The highest output power obtained from an laser diode pumped Nd: glass laser was reported by Fan. Using a spinning glass disc as the gain medium to overcome thermal problems, an output power of 550 mW was obtained for an absorbed power of 2 W. Basu and Byer carried out an analysis of the scalability of laser diode pumped Nd: glass lasers in the zig zag slab and rotating disc geometries. They predicted that for the rotating disc laser, up to 20 kW of average output power should be achievable.

Diode Pumped Stoichiometric Lasers

We consider the work carried out to date on stoichiometric laser materials. The stimulus for the research into these materials is essentially to reduce the size of DPSS lasers. Stoichiometric materials are compounds which contain the lasing species, rather than hosts into which the species is doped. Due to this, the concentration of active ions can be much larger in stoichiometric materials, resulting in a shorter absorption length for the pump radiation.

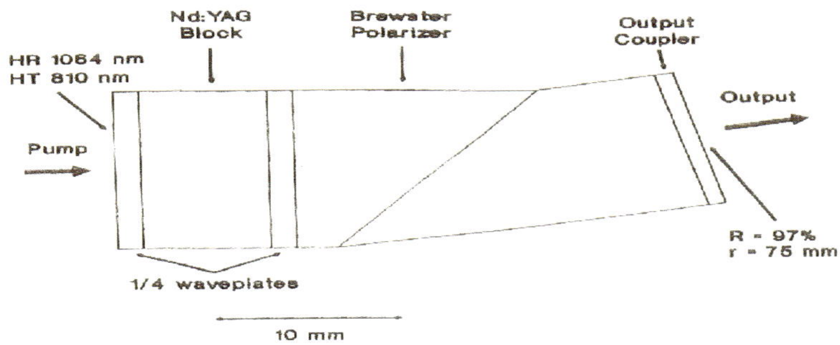

Figure: Monolithic twisted mode cavity Nd:YAG laser

Dixon et al. have used an interesting technique to couple the pump radiation into the LNP ($LiNdP_4O_{12}$) laser mode where the crystal is placed in very close proximity to the laser diode output facet. Due the very short absorption length of the LNP crystal, there is no need for collimating and focusing optics to be used, as is normally the case for DPSS lasers. Using a separate output coupler, laser operation was obtained at both 1047 nm (slope efficiency 33%) and 1317 nm (slope efficiency 10%). The maximum output powers obtained were 73.5 mW and 24 mW, respectively. Using a monolithic device, 28 mW output was obtained at 1317 nm. The authors believe that the slope efficiencies reported will increase as crystal growth and fabrication techniques are improved. If the axes of the wave plates are properly adjusted, a standing wave with an axially uniform energy density can be created. Spatial hole burning is thus eliminated.

Diode Pumped Waveguide Lasers

In addition to bulk lasers, there is much interest in the use of laser diodes to pump waveguide lasers. The use of guided wave structures instead of bulk devices is also of interest for diode pumping. One reason is the compatibility of guided wave devices with optical fiber systems. Another reason is that by guiding both the pump wave and the laser mode, higher pump densities and therefore gain compared to bulk devices can be achieved. The first report of a guided wave device was end pumped a multimode, Nd doped, silica based, glass fiber laser. The cavity was provided by either depositing reflector coating directly onto the end of the fiber or by external mirrors.

Wave guiding structures were also considered for a transverse pump geometry. Such a device may offer advantages such as small resonator construction and good transverse mode stability. Design calculations were performed for an light emitting diode array pumped, rectangular cross section waveguide fabricated of a number of Nd laser materials including Nd: YAG, and LNP. Such a device made of an LNP gain medium with a glass cladding was demonstrated using Ar^+ laser pumping.

Thin-disk Lasers

The *thin-disk laser* (sometimes called *thin-disc laser* or *active-mirror laser*) is a special kind of diode-pumped high-power solid-state laser, which was introduced in the 1990s by the group of Adolf Giesen at the University of Stuttgart, Germany.

The main difference from conventional rod lasers or slab lasers is the geometry of the gain medium: the laser crystal is a thin disk, where the thickness is considerably smaller than the laser beam diameter. The heat generated is extracted dominantly through one end face, i.e., in the longitudinal rather than in the transverse direction. The cooled end face has a dielectric coating which reflects both the laser radiation and the pump radiation.

The thin disk is also often called an *active mirror,* because it acts as a mirror with laser gain. Within the laser resonator, it can act as an end mirror or as a folding mirror. In the latter case, there are two double passes of the laser radiation per resonator round trip, so that the gain per round trip is doubled and the threshold pump power is reduced.

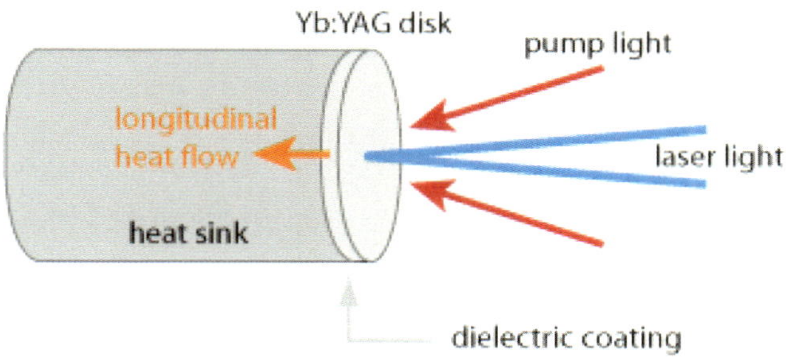

Figure: Setup of a thin disk laser head. The pump optics (not shown) are arranged for multiple passes of the pump radiation. The heat is extracted in the longitudinal direction, which minimizes thermal lensing effects

The thin-disk laser should not be confused with the *rotary disk laser*, where the gain medium is a quickly rotating disk, which is usually a few millimeters thick.

Multi-pass Pumping

The small thickness of the disk typically leads to inefficient pump absorption when only a single or double pass is used. This problem is normally solved by using a multi-pass pump arrangement, which can be made fairly compact when using a well-designed optical setup, typically containing a parabolic mirror and prism retro reflectors. Such arrangements easily allow one to arrange for e.g. 8 or 16 double passes of the pump radiation through the disk without excessively stringent requirements on the pump beam quality. Compared with high-power fiber lasers, thin-disk lasers have lower demands on the brightness of the pump diodes. However, the required brightness is higher than for some slab lasers and other side-pumped lasers.

The pump source of a thin-disk laser is usually based on high-power diode bars, either in fiber-coupled form or with free-space power delivery. A typical pump wavelength is 940 nm for Yb: YAG, whereas ytterbium-doped tungstate crystals can be more efficiently pumped near 981 nm.

Reduced Thermal Issues at High Output Powers; Power Scalability

Due to the small thickness of the disk (e.g. 100–200 μm for Yb: YAG), the temperature rise associated with the dissipated power is small. (It is not relevant that the density of heat generation is rather high, because the heat is generated very close to the heat sink.) In addition, the temperature gradients are dominantly in a direction perpendicular to the disk surface and thus cause only weak thermal lensing and depolarization loss. This allows for operation with very high beam quality due to the weak thermal beam distortions, and stable operation can be achieved over a wide range of pump powers.

A very important property arising from the thin-disk geometry is power scalability in a strict and meaningful sense. The scaling procedure is simple: for example, the output power can be doubled by applying twice the pump power to twice the area on the disk, while keeping the disk thickness and doping level constant. The laser resonator has to be modified so as to double the mode area in the disk. With this scaling procedure applied, the new design with twice the output power has unchanged peak optical intensities and a nearly unchanged maximum temperature in the disk (the latter essentially because the cooling area has also been doubled). As far as thermal lensing results from the temperature dependence of the refractive index, the dioptric power (inverse focal length) of the thermal lens is reduced to half of the original value, which just compensates the doubled sensitivity of the larger mode to focal length changes. The power has thus been scaled without increasing optical intensities, the magnitude of temperature rises, or thermal lens problems. Strictly, thermal lensing effects can be increased by the power scaling as far as they result from mechanical stress in the disk, but this effect can be kept small by keeping the disk thickness small. Another limitation arises from amplified spontaneous emission (ASE) in the transverse direction, which ultimately limits the gain achievable in the longitudinal direction, but becomes severe only at very high power levels with many kilowatts from a single disk. The use of a composite disk with an un doped part on top of the doped part can also strongly suppress ASE; at least for continuous-wave lasers, the ASE limit may be pushed to the order of 1 MW from a single disk.

Power scalability in a wide range is achieved even for passively mode-locked thin-disk lasers. Here, the doubling of the output power also involves doubling of the mode area on the SESAM, so that optical intensities in that device and also cooling issues are – contrary to naïve expectations – not limiting factors. Limitations rather arise from the challenge of implementing dispersion compensation at high power levels.

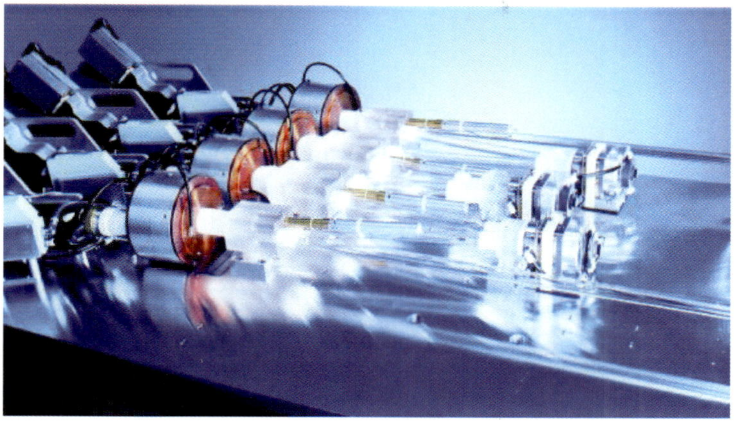

Figure: Setup of an 8-kW laser based on four thin-disk laser heads (seen on the left side)

So far, around 500 W of output power in a diffraction-limited continuous-wave beam have been obtained with a single disk, or around 1 kW using two disks. In multimode operation, more than 4 kW per disk are possible, and e.g. 16 kW using four disks in one resonator. Multimode thin-disk lasers with 16 kW output power are commercially available, and similar or even higher power levels in nearly diffraction-limited beam appear to be feasible. With mode-locked thin-disk lasers, over 240 W of average output power has been achieved.

Thin-disk Gain Media

The most often used gain medium for thin-disk lasers is Yb: YAG. Compared with Nd:YAG, it has a shorter emission wavelength (typically 1030 nm), a smaller quantum defect(reducing the dissipated power), a longer upper-state lifetime (improving energy storage for Q switching), and a larger gain bandwidth (e.g. for shorter pulses with mode locking). On the other hand, it is a quasi-three-level gain medium with significant reabsorption at the laser wavelength, and thus requires higher pump intensities. The thin-disk principle is well adapted to these parameters.

For broad wavelength tuning and for ultra short pulse generation, other ytterbium-doped gain media offer a still wider gain bandwidth. Examples are tungstate crystals (Yb: KGW, Yb: KYW, Yb: KLuW), Yb:$LaSc_3(BO_3)_4$ (Yb: LSB), Yb:$CaGdAlO_4$ (Yb: CALGO) and Yb:YVO_4. Particularly promising are novel sesquioxide materials such as Yb:Sc_2O_3, Yb:Lu_2O_3 and Yb:Y_2O_3, having excellent thermo-mechanical properties and a potential for very high output powers and high efficiencies. A slope efficiency of 80% has been demonstrated with Yb:Lu_2O_3. Research is under way concerning the use of titanium–sapphire disks, offering an extremely broad gain bandwidth for pulse generation in the sub-100-fs region.

Nd: YAG or Nd:YVO_4 may also be used in thin-disk lasers, e.g. when a wavelength of 1064 nm is required, or when the much smaller saturation energy of Nd:YVO_4 is relevant.

Generally, a high doping concentration is desirable for thin-disk gain media. This allows one to use a rather thin disk (and thus to minimize thermal effects) without arranging for too many passes of the pump radiation. Most ytterbium-doped gain media are quite favorable in this respect. Some materials, particularly laser glasses, which exhibit only weak pump absorption, are not well suited for the thin-disk geometry.

Spatial Hole Burning

An interesting consequence of the small disk thickness is that spatial hole burning usually cannot be avoided, even if a thin-disk laser is built with a ring resonator. (Note that due to the small ratio of thickness and beam radius, counter propagating waves in the disk always have a strong overlap, so that an interference pattern is generated even in a ring resonator.) Nevertheless, single-frequency operation is possible by using an appropriate wavelength filter (etalon) in the resonator. For passive mode locking, spatial hole burning in the thin disk distorts the shape of the gain spectrum, causing a variety of instabilities, but also allowing for significantly shorter pulses in the optimum range of parameters.

Q-switched Pulse Generation

Thin-disk lasers are well suited for generating high-energy nanosecond pulses with high beam quality, as required for, e.g., some kinds of laser material processing. The typically used gain medium Yb:YAG offers a significantly better energy storage (longer upper-state lifetime) compared with, e.g., Nd:YAG. A limiting factor, however, is the moderate gain (compared with that of an end-pumped rod laser), which makes it difficult to achieve very short (few-nanosecond) pulses with simple Q switching. The combination with cavity dumping is required to obtain short pulses.

Mode-locked High-power Thin-disk Lasers

Thin-disk lasers are particularly attractive for the generation of ultrashort pulses at very high power levels. In addition to the high-power capability, the main advantages in this context are:

- The ease of achieving diffraction-limited operation (which is a prerequisite for mode locking)
- The broad gain bandwidth of Yb:YAG (the so far most suitable gain medium for thin-disk lasers)

- The very small nonlinearity of a thin disk, which helps to avoid excessive nonlinear phase shifts despite the high intracavity peak intensities.

One of the initial challenges was to find a suitable mode-locking mechanism. Even though originally it was widely believed that passive mode locking with semiconductor saturable absorber mirrors (SESAMs) would not be possible at very high power levels – at least not without first developing special high-power SESAMs, possibly based on improved semiconductor materials, the author's research group at ETH Zürich demonstrated around 2000 that both thermal and non-thermal issues can be relatively easily managed even at very high power levels, if only the design parameters of the overall laser system (and not only of the SESAM) are properly chosen. In other words, SESAM damage then does not constitute a limiting factor for the power scaling of mode-locked thin-disk lasers; in fact, it was found that the mode-locked thin-disk laser is the first truly power-scalablefemtosecond laser. However, the design of such lasers involves a number of subtle issues, related to e.g. spatial hole burning and to dispersion compensation, and a trial-and-error approach not based on a solid understanding of various details is prone to fail, e.g. by not managing to suppress certain types of instabilities.

Thin-disk lasers have lead to the highest average output powers of far over 100 W from a mode-locked laser. Pulse energies of > 10 μJ combined with sub-picosecond pulse durations are possible, or 80 μJ in picosecond pulses. While typical pulse durations with Yb:YAG are around 700–800 fs, significantly shorter pulses are possible, e.g. with ytterbium-doped tungstate crystals such as Yb:KGW or Yb:KYW. Even with Yb:YAG, pulse durations around 200 fs have been demonstrated with Kerr lens mode locking.

Amplifiers for High Pulse Energies

Thin-disk laser heads can also be used for regenerative amplifiers. The relatively small gain of the thin disk can be compensated with a larger number of resonator round trips, even though this makes the amplifier more sensitive to optical losses and nonlinearities (e.g. in a Pockels cell). Therefore, it can be advantageous to arrange for multiple passes of the signal radiation through the disk in each resonator round trip. Substantial pulse energies are then possible even without using chirped-pulse amplification.

It is also possible to construct a purely multi pass amplifier (without a resonator and optical switch), but this approach limits the overall gain and requires a carefully optimized setup in order to preserve a high beam quality.

Nonlinear Frequency Conversion

High-power continuous-wave green lasers can be easily realized in the form of intracavity frequency-doubled thin-disk lasers. For Q-switched or mode-locked lasers, extracavity doubling is often more practical. Thin-disk lasers are also very interesting pump sources for optical parametric oscillators, amplifiers and generators. A high-power RGB source based on a mode-locked thin-disk laser has also been demonstrated. The high peak power of mode-locked thin-disk lasers allows for efficient nonlinear frequency conversion with critical phase matching in LBO crystals, thus not requiring temperature-stabilized crystal ovens for most or all conversion stages.

Further Development of the Thin-disk Concept

The thin-disk laser concept allows for further variations. For example, side pumping of the disk may allow for even higher output powers while reducing the requirements on the pump beam quality. This approach, developed at the Lawrence Livermore National Laboratory, is based on a composite laser crystal. An un doped YAG disk, which is bonded to a Yb: YAG disk, brings various benefits: It reduces the beam quality requirements for the pump source, reduces the tendency for transverse amplified spontaneous emission (ASE) and parasitic lasing in large disks, increases the mechanical strength and may improve the cooling. The side-pumped concept may thus allow scaling to much higher powers, even though not necessarily with diffraction-limited beam quality. An interesting option is to use a composite ceramic gain medium which is ytterbium doped only in the center region and not in the outside regions, which are used only to deliver the pump power.

It is also interesting to develop cryogenic thin-disk lasers, as cryogenic cooling greatly reduces thermal effects at high power levels.

Competition with Fiber Lasers

Thin-disk lasers are currently facing fierce competition from high-power fiber lasers and amplifiers. In continuous-wave operation, these can currently deliver even higher powers in close to diffraction-limited beams. Within the next few years, both thin-disk lasers and fiber lasers are expected to show significant further progress, and it is currently not clear which technology will acquire the larger market share.

In the domain of ultra-short pulse generation, fiber amplifier systems based on chirped-pulse amplification allow one to reach even higher average powers and shorter pulse durationsthan thin-disk lasers can generate without amplification. However, such systems are relatively complex, and thin-disk lasers will probably maintain superiority for the direct generation of pulses with high energies, particularly when high pulse quality (concerning temporal shape, low chirp, and stable linear polarization) are required. A key issue in this context is that both the small disk thickness and the larger beam diameter on the disk lead to a nonlinearity which, compared with that of a fiber laser, is smaller by many orders of magnitude, and thus in many cases allows the use of much simpler techniques.

A detailed comparison of thin-disk versus fiber lasers is complex and has to take into account many aspects which depend on the particular application. For example, issues such as emission bandwidth, pulse quality and stability of polarization state can be essential in some cases but insignificant in others.

Semiconductor Disk Lasers

The thin-disk geometry is also used in vertical external cavity surface-emitting lasers (VECSELs), a kind of semiconductor lasers. In that case, multiple passes of the pump are usually not required due to the strong absorption of semiconductor materials. However, the concept with multiple pump passes has recently also been applied to such semiconductor lasers, where it allows for a reduced quantum defect and thus for reduced heating and potentially higher powers. To date it is not clear whether this will lead to more efficient and practical lasers; tens of watts of output power have so far been reached with the original concept, not requiring multiple pump passes. Such lasers are actually very interesting, partly because they can be developed at different wavelengths, e.g. for blue light generation with intracavity frequency doubling at very high output power levels.

Er: YAG Laser

Er: YAG lasers are solid-state lasers that have erbium-doped yttrium aluminium garnet as a lasing medium.

The erbium: YAG laser is a 2940-nm ablative laser that targets intracellular water like the CO_2 laser does; however, its affinity for intracellular water is 10 times greater than that of carbon dioxide laser. This greater affinity allows for a cleaner target cell ablation with less thermal collateral cell damage. Collagen shrinking is reduced with the erbium, resulting in decreased effect on rhytids. Yet the erythema and complications are greatly diminished compared with the CO_2 laser due to the decreased collateral thermal necrosis.

The zone of thermal necrosis can be modified from 20-micron to 200-micron ablation based on the pulse width chosen. Even with variable pulse widths the side-effect profile and recovery period are still more favorable than with CO_2 resurfacing.

Care must be taken with the erbium laser because the cellular char that is present after CO_2 resurfacing is not present with the erbium. The char layer can limit the depth of further dermal ablation due to the decreased water content in this layer. However, the erbium laser affinity for water is so great and the ablation so clean that deeper ablation is possible without the visible warning provided by the CO_2 laser.

Laser Properties	
Laser type	Solid
Pump source	Laser diode
Operating wavelength	1.53-1.56 μm

Physical and Chemical Properties	
Chemical formula	$Er^{3+}:Y_3Al_5O_{12}$
Molecular weight	593.7 g mol^{-1}
Crystal structure	Cubic
Melting point	1965°C
Mohs hardness	8.25
Thermal conductivity	11.2 W m^{-1} K^{-1}
Specific heat capacity	0.59 J g^{-1} K^{-1}
Young's modulus	335 GPa

Applications

Er: YAG lasers have been widely used for laser resurfacing of human skin that includes melasma, deep rhytides and acne scarring. The energy of Er: YAG lasers is also absorbed well by hydroxyapatite in addition to water, and hence they are suitable for cutting bone as well as soft tissues.

They are effective for treating tooth decay atraumatically, without the use of local anesthetia to numb the tooth. They eliminate the vibration of the dental drill to avoid the risk of causing micro fractures in the tooth. However, the laser energy provides a sedative effect on the nerve when used at low power settings thereby creating a sensation of pain in the tooth.

Other major applications of Er: YAG laser include:

- Dentistry
- Removal of warts
- Drug delivery.

Ruby Laser

A ruby laser is a solid-state laser that uses the synthetic ruby crystal as its laser medium. Ruby laser is the first successful laser developed by Maiman in 1960.

Ruby laser is one of the few solid-state lasers that produce visible light. It emits deep red light of wavelength 694.3 nm.

Construction of Ruby Laser

A ruby laser consists of three important elements: laser medium, the pump source, and the optical resonator.

Laser Medium or Gain Medium In Ruby Laser

In a ruby laser, a single crystal of ruby ($Al_2O_3 : Cr^{3+}$) in the form of cylinder acts as a laser medium or active medium. The laser medium (ruby) in the ruby laser is made of the host of sapphire (Al_2O_3) which is doped with small amounts of chromium ions (Cr^{3+}). The ruby has good thermal properties.

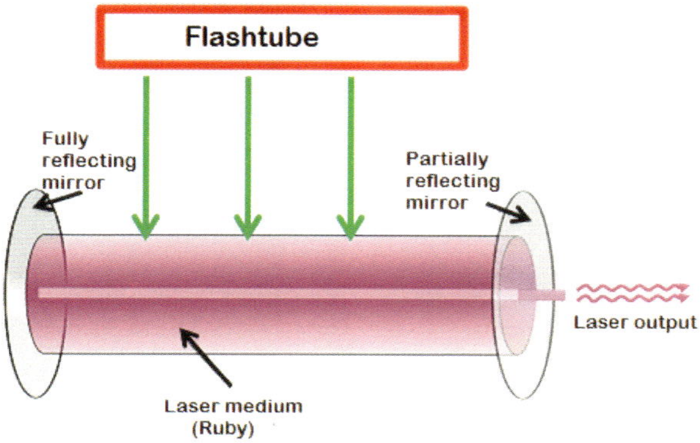

Pump Source or Energy Source in Ruby Laser

The pump source is the element of a ruby laser system that provides energy to the laser medium. In a ruby laser, population inversion is required to achieve laser emission. Population inversion is the process of achieving the greater population of higher energy state than the lower energy state. In order to achieve population inversion, we need to supply energy to the laser medium (ruby).

In a ruby laser, we use flashtube as the energy source or pump source. The flashtube supplies energy to the laser medium (ruby). When lower energy state electrons in the laser medium gain sufficient energy from the flashtube, they jump into the higher energy state or excited state.

Optical Resonator

The ends of the cylindrical ruby rod are flat and parallel. The cylindrical ruby rod is placed between two mirrors. The optical coating is applied to both the mirrors. The process of depositing thin layers of metals on glass substrates to make mirror surfaces is called silvering. Each mirror is coated or silvered differently.

At one end of the rod, the mirror is fully silvered whereas, at another end, the mirror is partially silvered.

The fully silvered mirror will completely reflect the light whereas the partially silvered mirror will reflect most part of the light but allows a small portion of light through it to produce output laser light.

Working

Ruby is a three level laser system. Suppose there are three levels E1, E2 and (E3 & E4). E1 is the ground level, E2 is the metastable level, E3 and E4 are the bands. E3 & E4 are considered as only one level because they are very closed to each other.

Pumping: The ruby crystal is placed inside a xenon flash lamp and the flash lamp is connected to a capacitor which discharges a few thousand joules of energy in a few milliseconds. A part of this energy is absorbed by chromium ions in the ground state. Thus optical pumping raises the chromium ions to energy levels inside the bands E3 and E4. This process is called stimulated absorption. The transition to bands E3 and E4 are caused by absorption of radiations corresponding to wavelengths approximately 6600 angstroms and 4000 angstroms respectively. The levels inside the bands E3 and E4 are also known as pumping levels.

Achievement of population inversion: Cr^{3+} ions in the excited state loose a part of their energy during interaction with crystal lattice and decay to the metastable state E2. Thus, the transition from excited states to metastable state is non-radiative transition or in other words there is no emission of photons. As E2 is a metastable state, so chromium ions will stay there for longer time. Hence, the number of chromium ions goes on increasing in E2 state, while due to pumping, the number in the ground state E1 goes on decreasing. As a result, the number of chromium ions become more in excited state (metastable state) as compared to ground state E1. Hence, the population inversionis achieved between states E2 and E1.

Achievement of laser: Few of the chromium ions will come back from E2 to E1 by the process of spontaneous emission by emitting photons. The wavelength of a photon is 6943 Å. This photon travels through the ruby rod and if it is moving in a direction parallel to the axis of the crystal, then it is reflected to and fro by the silvered ends of the ruby rod until it stimulates the other excited ions and cause it to emit a fresh photon in phase with the stimulating photon. Thus, the reflections will result in stimulated emission and it will result in the amplification of the stimulated emitting photons. This stimulated emission is the laser transition.

The two stimulated emitted photons will knock out more photons by stimulating the chromium ions and their total number will be four and so on. This process is repeated again and again, thus photons multiply. When the photon beam become sufficiently intense, then a very powerful and narrow beam of red light of wavelength 6943 Å emerges through the partially silvered end of the ruby crystal.

In the energy level diagram, E2 is the upper laser level and E1 is the lower laser level because laser beam is achieved in between these levels. Thus, the ruby laser fits into the definition of three level laser system.

Output: The output wavelength of ruby laser is 6943 Å and output power is 10 raise to power 4 to 10 raise to power 6 watts and it is in the form of pulses.

Spiking in Ruby Laser

The terminus of laser action is the ground state E1 in ruby laser. Therefore it is difficult to maintain the population inversion. This will result in the depletion of upper laser level E2 population (due to stimulated emission) more rapidly than it can be restored by the flash light that is optical pumping source. The laser emission is made up of spikes of high intensity emissions. This phenomenon is called spiking of the laser.

After the depletion of E2 state, the laser action ceases for a few microseconds. Since the flash lamp is still active, it again pumps the ground state chromium ions to upper level and again laser action begins. A series of such pulses is produced until the intensity of the flash light has fallen to such a level that it can no linger rebuild the necessary population inversion. So the output laser will be in the form of pulse in ruby laser or in other words.

Energy Level Diagram for Ruby Laser

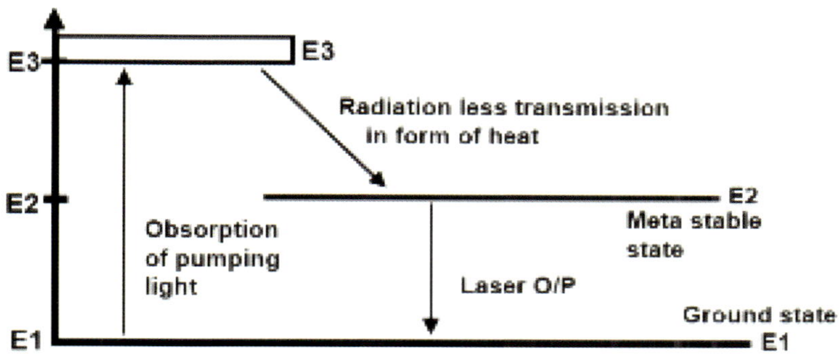

The above three level energy diagram show that in ruby lasers the absorption occurs in a rather broad range in the green part of the spectrum. This makes raise the electrons from ground state E1 to the band of level E3 higher than E1. At E3 these excited levels are highly unstable and so the electrons decays rapidly to the level of E2. This transition occurs with energy difference (E1 – E2) given up as heat (radiation less transmission). The level E2 is very important for stimulated emission process and is known as metastable state. Electrons in this level have an average life time of about 5m.s before they fall to ground state. After this the population inversion can be established between E2 and E1. The population inversion is obtained by optical pumping of the ruby rod with a flash lamp. A common type of the flash lamp is a glass tube wrapped around the ruby rod and filled with xenon gas. When the flash lamp intensity becomes large enough to create population inversion, then stimulated emission from the metastable level to the ground level occurs which result in the laser output. Once the population inversion begins, the Metastable level is depopulated very quickly. Thus the laser output consists of an intense spike lasting from a few Nano sec to μsec. after stimulated emission spike, population inversion builds up again and a 2nd spike results. This process continues as long as continues as long as the flash lamp intensity is enough to create the population inversion.

Drawbacks of Ruby Laser

1. As the terminus of laser action is the ground state, it is difficult to maintain the population inversion. This fact results in ruby laser's low efficiency.
2. The ruby laser requires high power pumping source.
3. The laser output is not continuous but occurs in the form of pulses of microsecond duration.
4. The defects due to crystalline imperfection are also present in ruby laser.

Uses of Ruby Laser

1. Ruby laser has very high output power of the order of 10^4 – 10^6watts. It has wavelength of 6943 Angstroms.
2. Ruby lasers are used for holography, industrial cutting and welding.

Solid-state Dye Lasers

The use of solid matrices containing laser dyes is an attractive alternative to the conventional liquid dye solutions. The first solid-state dye lasers were reported in the late 1960s by Soffer and Mcfarland and Peterson and Snavely. They demonstrated stimulated emission from polymeric matrices doped with organic dyes. However work on solid-state dye lasers was not pursued for over a decade due to low lasing efficiencies and fast photo degradation of the dye. In recent years, significant breakthroughs have been achieved in the development of practical tunable solid-state dye lasers. The solid-state dye lasers have several advantages over the conventional liquid dye lasers. Along with the ease of handling, these lasers possess commercial advantage because of low cost of production and the safety of operation. Other technical advantages are compactness, manageability, versatility, lack of flammability and lack of toxicity. The flow fluctuation and solvent evaporation is considerably reduced in the case of solid-state dye lasers.

Types of Host Materials

There are a number of materials which have been used as solid hosts for laser dyes such as polymers, porous glasses, organically modified silicates or silicate nano-composites, polycom glass (combination of polymer and sol–gel). Solid-state dye lasers based on polymeric and sol–gel glass have been reported extensively in literature. In general, the solid host materials suitable for use in solid-state dye lasers should have the following requirements. They should be highly transparent to the pump laser and dye laser wavelengths. They should be hard enoughto undergo mechanical treatment suchas diamond cutting, grinding and polishing as required by any optical material and the materials should possess good thermal stability over a wide range of temperatures. Polymeric matrices have some important advantages over other host materials because they are simple to prepare and cheap to produce. In general, the polymers used are polyacrylics, polyurethanes (PUA), polycarbonates, etc. In polyacrylic class, the commonly used polymeric hosts are polymethylmethacrylate (PMMA). They possess high optical homogeneity, their chemical composition can be controlled at the formation stage and they can be treated mechanically. These polymers are highly transparent in the visible region of the electromagnetic spectrum. Polymer based systems have limitations such as a low damage threshold of the host material and limited lifetimes. Polymers are poor thermal conductors and have large time of heat dissipation τ from the focused pump laser spot to low heat boundary (for PMMA, $\tau = 0{:}5$ s).

A large number of laser dyes are insoluble in PMMA therefore low molecular weight additives such as ethyl alcohol, ethylene glycol, ether, polyether are added to PMMA to form modified PMMA known as MPMMA. Low molecular weight additives together with monomer purification can also lead to increased laser damage threshold. P(MMA: HEMA) is a copolymer of heterogenous mixtures of monomers such as methylmethacrylate (MMA) and 2-hydroxyethyl methacrylate (HEMA). When this mixture is polymerised it provides enhanced crosslinking and compatibility of dye with macromolecule. HEMA provides enhanced solvation for the dye molecule. Costella et al. have reported improved laser operation with P(MMA:HEMA) in some specific cases. More recently solid copolymers of methyl methacrylate (MMA) and trifluoromethyl methacrylate (TFMA) along with crosslinking monomer ethylene glycol dimethacrylate (EGDMA) have been reported. Th is matrix was found suitable for the highly efficient and photo-stable pyrromethene 567 dye as a dopant. Copolymers of methyl methacrylate (MMA) withTFMA lead to materials, which have mechanical and optical properties, suited to efficient laser systems. These new solid materials show lasing efficiencies up to 18% when transversely pumped at 534 nm using Nd:KGW laser with good photo stabilities. Co-polymers of MMA (methylmethacrylate) and pentacrythitrol triacrylate (PETA) when pumped with the green line of copper vapour laser produced 290 mW of average power at 1 KHz repetition rate after which the laser power decreased. Polymer filled porous glass impregnated with dye molecules has also been used as an active medium. Th e thermo-optical and mechanical parameters are determined by a matrix of silica whose thermo optical constants are better by two orders of magnitude than the polymers. However these materials have optical in homogeneities, which can affect laser performance. Rhoda mine 6G dye doped in polymer and porous glass composition can withstand pump pulses of 50 ns with energies of 7–10 J/cm^2. Such systems are used on-board, in vacuum under low temperature where the dye flow system operation becomes a problem.

Recently, a narrowband laser with dye doped polymer as the gain medium with a line width of ~ 0.011 cm^1 (350 MHz) at FWHM has been reported by Duarte. Tunable lasing from a dye laser in the near infrared region 1.1 mm using polymethine dye in a polyurethane matrix pumped by 1.06 mm was reported. The conversion efficiency was 43% and tuning range ~ 520.6 cm^1 (63 nm).

Photo Stability and Quantum Efficiency of SSDL Materials

The photo stability of the dye doped solid materials depends on the inter-molecular and intra-molecular interactions of the dye molecule with surrounding chemically active molecules such as the polymer macromolecule end groups, unreacted monomers, absorbed atmospheric oxy- gen molecules and another dye molecule. In general there are three accepted reasons for degradation of the dye molecule:

1. Photo deactivation from the excited state caused by chemical oxidation reaction.
2. Formation of dimers that absorbs irradiation without any fluorescence.
3. Thermal destruction.

Dye doped solid-state laser materials have low quantum efficiency and a limited useful operation time. The low quantum efficiency is mainly due to photo-destruction of dye molecules as mentioned above. The mobility and the concentration of the dye molecules are the important factors that decide the probability of the photo deactivation reaction. Different methods are used to decrease the mobility of the dye molecules and to reduce the rate of destructive collisions and diffusion out of the active area. The mobility of the dye molecule is minimized by the following methods. The dye can be doped in copolymer host e.g.: P (MMA: HEMA). Due to crosslinking in the copolymer the mobility of the dye molecule is reduced. Another method is filling free volumes in the polymer host with low molecular weight additives. It is important that additives do not form any chemical bond with dye or macromolecules so that the optical properties of the material do not get affected. These additives also improve the thermal conductivity of the material and help cool the lasing material. The additives are smaller in size than the dye molecules and polymer macromole- cules. Hence the additives fill the polymer free volumes more easily and also prevent the dye from diffusing into the thermal region. Working with polymer matrices that contain low molecular weight additives one can achieve considerably longer operational lifetimes than has been reported using similar matrices without additives. One such additive is di phenylthiourea. It has been reported that addition of di phenylthiourea in copolymers of PMMA increased the photo stability of the copolymer samples containing laser dyes like rhodamine and coumarin. By doping dye in a polymer host with controlled crosslinking the photo stability and the efficiency were improved for the pyromethene 567 (PM 567) in poly trifluromethyl methacrylate (TFMA) with cross- linking monomer ethylene glycol di methacrylate (EGDMA).

Another reason for the degradation of dye molecules is thermal destruction of the dye because of the low thermal conductivity of the polymer host. Since the polymer host is transparent to the pump radiation, the excited dye molecule heats up the host through non-radiative thermal relaxation. The polymers have poor thermal conductivity. Also a large time of heat dissipation from the focused pump laser spot to low heat boundary, limits the repetition rate of the dye laser to few hertz. This accumulated heat increases the mobility of the dye and increases the probability of photo deactivation reaction with the surrounding chemically active molecules. Hence the direct thermal destruction of the material has insignificant role. To overcome this problem, the dye molecules are covalently linked to the polymeric chains so that the additional channels for the dissipation of the excess heat energy are provided.

Polymers have been developed with higher laser damage resistance, comparable with inorganic glasses and crystals. The single shot damage resistance of MPMMA was higher than 13 j/cm^2 for pump spot of 1 mm diameter. Single shot damage is caused either by inclusions or micro-structures and the multiple shot damage is intimately related to the viscoelastic properties of the host. By controlling the inter and intra- molecular interactions of dye with polymer host the conversion efficiency and the critical number of pump pulses before the photo bleaching sets can be improved. The conversion efficiency can also be improved by choosing dye with large volume anion substitutes. This shifts the equilibrium of monomer to dimer reversible reaction towards the monomer side and improves the quantum yield and conversion efficiency. The experiments with Rhoda mine 6G chloride and Rhoda mine 6G per- chlorate show that the latter has large quantum yield and hence large conversion efficiency. The conversion efficiency can be further improved by adding Low Molecular Additives (LMA) to the polymer, which enhances the dielectric con-stant of the polymer and facilitate the formation of expanded ionic pairs. The efficiency of the PMMA based Rhoda mine 6G chloride dye was improved from 14% to 28% by the addition of ethyl alcohol–carbonic acid ether mixture into PMMA.

Dyes used in Solid-state Dye Lasers

The dyes to be doped in the polymeric host are chosen depending upon the desired wavelength of operation, efficiency and photo stability of the dye molecule. Coumarin and rhodamine dyes are frequently used in solid-state dye lasers. Coumarins are laser dyes, which show laser emission in the blue–green region of the spectrum. The presence of a hydroxy or amino group at the 7-position has been found to be necessary for the merocyanine chromophore to exhibit lasing action. Replacement of the two amino hydrogens by various alkyl groups has been reported to increase their lasing efficiency.

Introduction of a phenyl group at the 3-position has also been found to increase the lasing efficiency and a number of 7- N,N-diallylamino-3-phenyl coumarins have been synthesized and their lasing action reported.

Coumarin 10

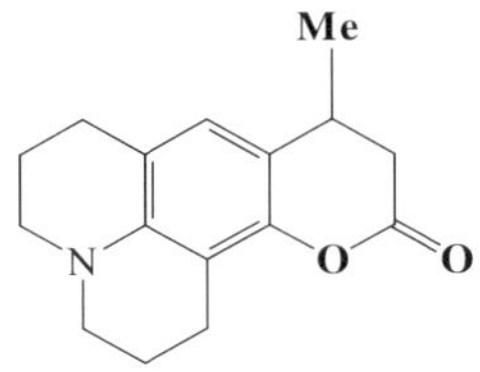

Absorption Max 389 nm   Lasing Max 480 nm

Coumarin 7

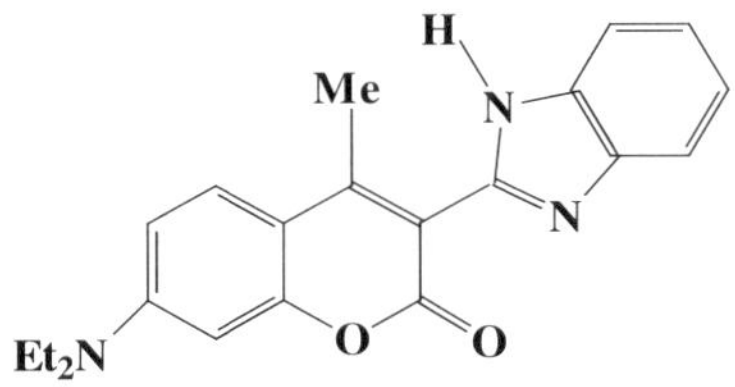

Absorption Max 436 nm   Lasing max 573 nm

The structure of rhodamine 6G, which is more often used as a laser dye is shown above. Several other derivatives of rhodamines are also frequently used depending on the requirements for lasing.

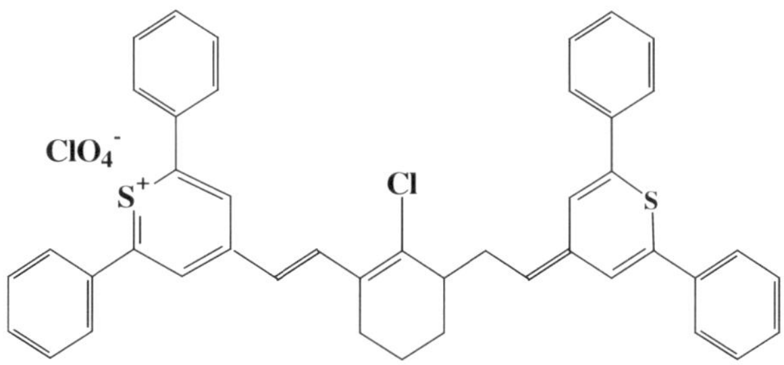

Thiopyrlocarbocyanine   Absorption Max 1064nm

The above dye thiopyrlocarbocyanine is used more commonly in polyurethane host. The maximum in its main absorption band in polyurethane lies at 1.079 mm. The thiopyrylotricarbocyanine dye has a relatively large excited state relaxation time in liquid solutions of the order of 50 ps.

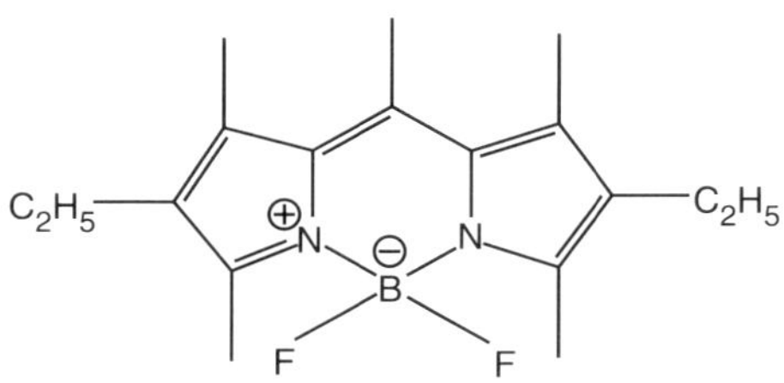

Pyrromethene567 (PM 567)   Absorption Max 561 nm

Pyrromethene dyes of the above structure are also used for many related applications. Pyrromethene BF_2 complexes are a class of laser dyes, which have been synthesized during late 1980s and early 1990s. In certain cases the pyrromethene and the polymethine laser dyes have been found to give better results than the rhodamines and the coumarin dyes. The pyrromethene dyes show reduced triplet–triplet absorption over the fluorescence and lasing spectral region but at the same time exhibit a high fluorescence yield. These dyes show laser emission over the spectral region from the green–yellow to red. These dyes have demonstrated lasing with good performance when incorporated into solid hosts. Oxazine dyes such as cresylviolet, nile blue, etc. are also used as laser dyes. These oxazine dyes are cationic dyes.

Aggregation is well known to take place for many cationic dyes in water. Dye aggregates are an integral component of many imaging and light conversion application. The aggregates of the oxazine laser dye have been a focus of research owing to the potential industrial applications of the aggregates strong absorption bands in the visible oxazine dyes are used as laser dyes because of their long term stability and high quantum yield.

Different Types of Solid-state Dye Lasers

The pump lasers employed in the solid-state dye lasers depend upon the laser dye doped in the host. For the dyes such as rhodamine 6G, rhodamineB, pyrromethene, perylene orange dyes operating in the visible range 550–650 nm, the most widely used pump source is second harmonic (532 nm) of Nd: YAG laser. For coumarin class of dyes the pump lasers used are Excimer, third harmonics of Nd: YAG and Nitrogen laser.

In a recent communication polymethine dye doped in a polyurethane matrix was pumped by Nd:YAG laser operating at 1.06 mm. Operational lifetimes of 106 shots at a pump energy of 270 mJ/ pulse with output energies >100 mJ have been achieved. Using dye doped polyurethane acrylate (PUA) prepared by radical photo polymerization, laser efficiency of 30% was obtained in a hemi-spherical resonator. Dual wavelength operation has been achieved from a single rod doped with two dyes. A mixture of pyrromethene 580 (PM580) and pyrromethene 650 (PM650) doped in high temperature plastic or Excilitet (HTP) in the ratio 10:1 gave dual wavelength emission centered at 552 and 616 nm. High repetition rate solid-state laser operating at a repetition rate of 25 KHz with a broad band output has been reported. Lasing was obtained at 4.0 W output for 2 h.

Single mode operation at line widths of B0.001 cm^1 has been obtained in solid-state dye lasers. A short multiple prism near grazing incidence cavity with Rh6G dye doped in modified PMMA was used as the gain medium. Solid-state dye lasers withan active medium of polymer filled porous glass impregnated withdye molecules have been reported. The thermo optical and mechanical parameters are determined by a matrix of silica whose thermo optic constants are better by two orders of magnitude than that of polymers. The conversion efficiencies using Rh6G dye were 50%. The polymers have also been used in the integrated compact distributed feedback laser (DFB) pumped by nanosecond Nd: YAG laser, using PMMA waveguide and the mixture of laser dye Rh6G and the saturable absorber 3,30 – Diethyloxadicarbocyanine Iodide (DODCI), generating a series of spikes of 20–30 ps and later compressed to 9 ps. Recently high repetition rate polymeric solid-state dye lasers pumped by copper vapour laser with the operation time of 8 h (B10^8 pulses) have been reported. The phenalemine 512 dye doped with 2.5 mmol/l was tried out in a transverse pumping geometry. This resonator geometry improves the useful lifetime of the host material.

Fiber Laser

Fiber lasers are usually meant to be lasers with optical fibers as gain media, although some lasers with a semiconductor gain medium (a semiconductor optical amplifier) and a fiber resonator have also been called fiber lasers (or *semiconductor fiber lasers*). Also, devices containing some kind of laser (e.g., a fiber-coupled laser diodes) and a fiber amplifier are often called fiber lasers (or fiber laser systems).

In most cases, the gain medium is a fiber doped with rare earth ions such as erbium (Er^{3+}), neodymium (Nd^{3+}), ytterbium (Yb^{3+}), thulium (Tm^{3+}), or praseodymium (Pr^{3+}), and one or several fiber-coupled laser diodes are used for pumping. Therefore, most fiber lasers are *diode-pumped lasers*. Although the gain media fiber lasers are similar to those of solid-state bulk lasers, the wave guiding effect and the small effective mode area usually lead to substantially different properties of the lasers. For example, they often operate with much higher laser gain and resonator losses. See also the article on fiber lasers versus bulk lasers.

Fiber Laser Resonators

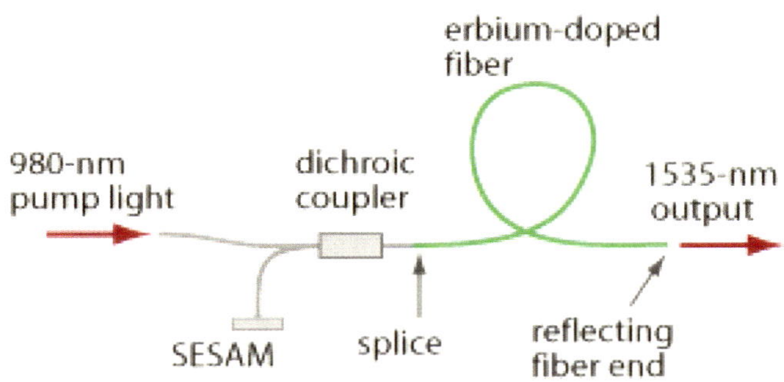

Figure: A simple erbium-doped femtosecond laser, where the Fresnel reflection from a fiber end is used for output coupling

In order to form a laser resonator with fibers, one either needs some kind of reflector (mirror) to form a linear resonator, or one builds a fiber ring laser. Various types of mirrors are used in linear fiber laser resonators:

- In simple laboratory setups, ordinary dielectric mirrors can be butted to the perpendicularly cleaved fiber ends. This approach, however, is not very practical for mass fabrication and not very durable either.
- The Fresnel reflection from a bare fiber end face is often sufficient for the output coupler of a fiber laser. Figure above shows an example.
- It is also possible to deposit dielectric coatings directly on fiber ends, usually with some evaporation method. Such coatings can be used to realize reflectivities in a wide range.
- For commercial products, it is common to use fiber Bragg gratings, made either directly in the doped fiber, or in an un doped fiber which is spliced to the active fiber. Figure below shows a distributed Bragg reflector laser (DBR laser) with two fiber gratings, but there are also distributed feedback lasers with a single grating in doped fiber, with a phase shift in the middle.

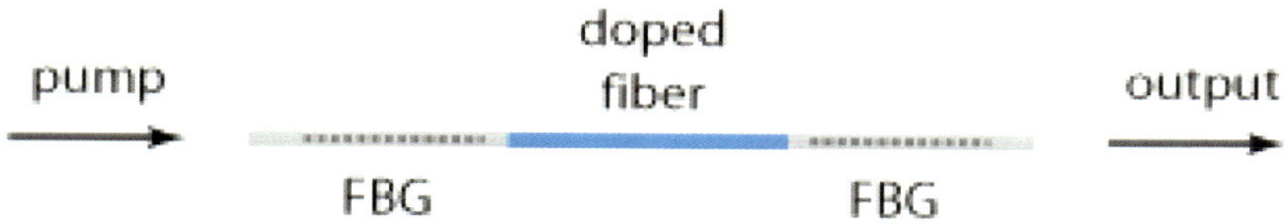

Figure: Short DBR fiber laser for narrow-line width emission

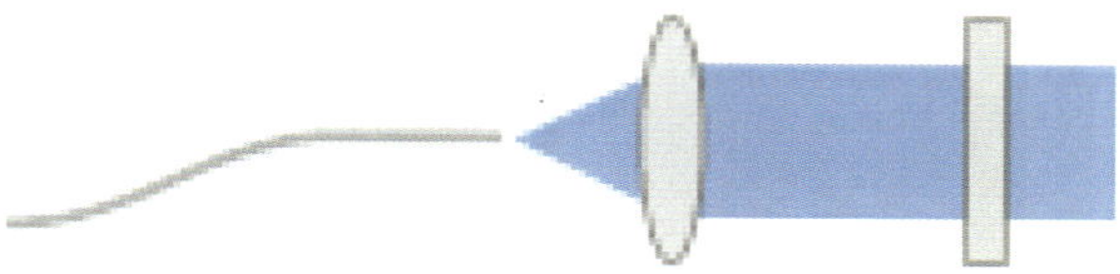

Figure: End reflector with lens and mirror.

- A better power-handling capability is achieved by collimating the light exiting the fiber with a lens and reflecting it back with a dielectric mirror. The intensities on the mirror are then greatly reduced due to the much larger beam area. However, slight misalignment can cause substantial reflection losses, and the additional Fresnel reflection at the fiber end can introduce filter effects and the like. The latter effects can be suppressed by using angle-cleaved fiber ends, which however introduce polarization-dependent losses.

Figure: Fiber loop mirror

- Another option is to form a fiber loop mirror, based on a fiber coupler (e.g. with 50:50 splitting ratio) and some piece of passive fiber.

Most fiber lasers are pumped with one or several fiber-coupled diode lasers. The pump light may be coupled directly into the core, or in high-power into a larger pump cladding (→ *double-clad fibers*).

There are many different kinds of fiber lasers, some of which are discussed in the following.

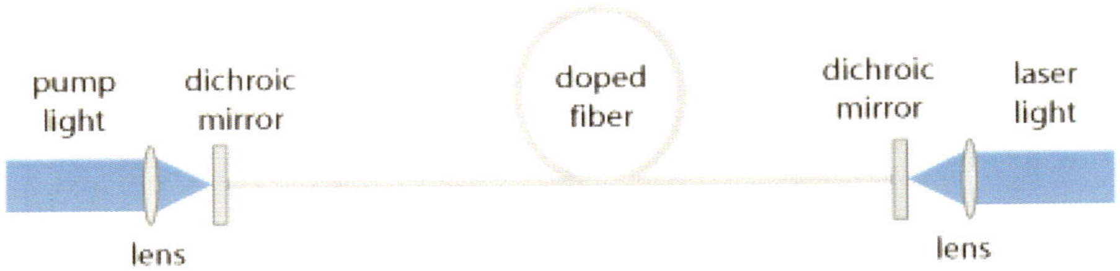

Figure: Setup of a simple fiber laser. Pump light is launched from the left-hand side through a dichroic mirror into the core of the doped fiber. The generated laser light is extracted on the right-hand side

High-power Fiber Lasers

Whereas the first fiber lasers could deliver only a few milliwatts of output power, there are now high-power fiber lasers with output powers of hundreds of watts, sometimes even several kilowatts from a single fiber. This potential arises from a high surface-to-volume ratio (avoiding excessive heating) and the guiding effect, which avoids thermo-optical problems even under conditions of significant heating.

Up-conversion Fiber Lasers

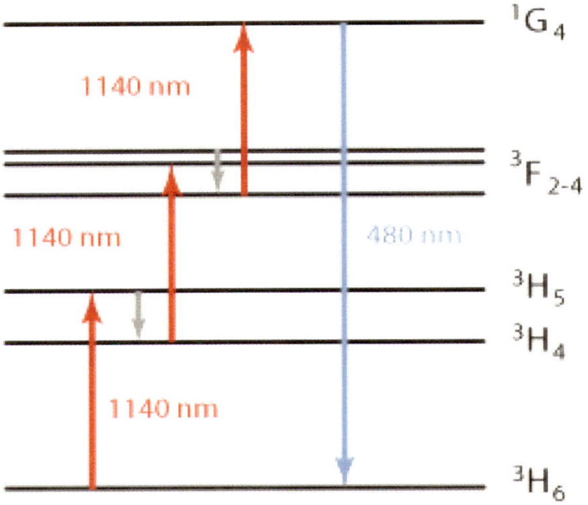

Figure: Level scheme of thulium (Tm^{3+}) ions in ZBLAN fiber, showing how excitation with an 1140-nm laser can lead to blue fluorescence and laser emission

The fiber laser concept is most suitable for the realization of up conversion lasers, as these often have to operate on relatively "difficult" laser transitions, requiring high pump intensities. In a fiber laser, such high pump intensities can be easily maintained over a long length, so that the gain efficiency achievable often makes it easy to operate even on low-gain transitions.

In most cases, silica glass is not suitable for up conversion fiber lasers, because the up conversion scheme requires relatively long lifetimes of intermediate electronic levels, and such lifetimes are often very small in silica fibers due to the relatively large phonon energy of silica glass (→ *multi-phonon transitions*). Therefore, one mostly uses certain heavy-metal fluoride fibers such as ZBLAN (a fluorozirconate) with low phonon energies.

The probably most popular up conversion fiber lasers are based on thulium-doped fibers for blue light generation, praseodymium-doped lasers (possibly with ytterbium co doping) for red, orange, green or blue output, and green erbium-doped lasers.

Narrow-line Width Fiber Lasers

Fiber lasers can be constructed to operate on a single longitudinal mode (→ *single-frequency lasers, single-mode operation*) with a very narrow line width of a few kilohertz or even below 1 kHz. In order to achieve long-term stable single-frequency operation without excessive requirements concerning temperature stability, one usually has to keep the laser resonator relatively short (e.g. of the order of 5 cm), even though a longer resonator would in principle allow for even lower phase noise and a correspondingly smaller line width. The fiber ends have narrow-bandwidth fiber Bragg gratings (→ *distributed Bragg reflector lasers, DBR fiber lasers*), selecting a single resonator mode. Typical output powers are a few milliwatts to some tens of milliwatts, although single-frequency fiber lasers with up to roughly 1 W output power have also been demonstrated.

An extreme form is the distributed-feedback laser (DFB laser), where the whole laser resonator is contained in a fiber Bragg grating with a phase shift in the middle. Here, the resonator is fairly short, which can compromise the output power and line width, but single-frequency operation is very stable.

Of course, further amplification to much higher power levels in a fiber amplifier is possible.

Q-switched Fiber Lasers

With various methods of active or passive Q switching, fiber lasers can be used for generating pulses with durations which are typically between tens and hundreds of nanoseconds. The pulse energy achievable with large mode area fiber scan be several millijoules, in extreme cases tens of millijoules, and is essentially limited by the saturation energy (even for large mode area fibers) and by the damage threshold (the latter particularly for shorter pulses). All-fiber setups (not containing any free-space optics) are quite limited in terms of the achievable pulse energy, as they can normally not be realized with large mode area fibers and effective Q switches.

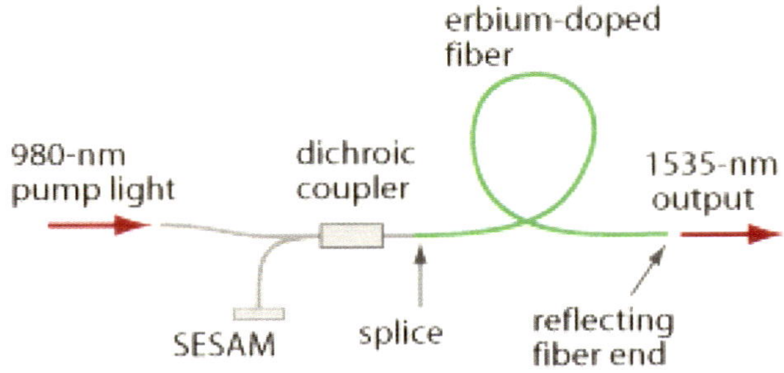

Figure: Simple Q-switched fiber laser. The setup looks exactly the same as that of a mode-locked laser as shown, but the SESAM parameters are different

Due to the high laser gain, the details of Q switching a fiber laser are often qualitatively different from those of a bulk laser, and more complicated. One often obtains a temporal sub-structure with multiple sharp spikes, and there is a possibility of producing Q-switched pulses with a duration well below the (typically long) resonator round-trip time.

Mode-locked Fiber Lasers

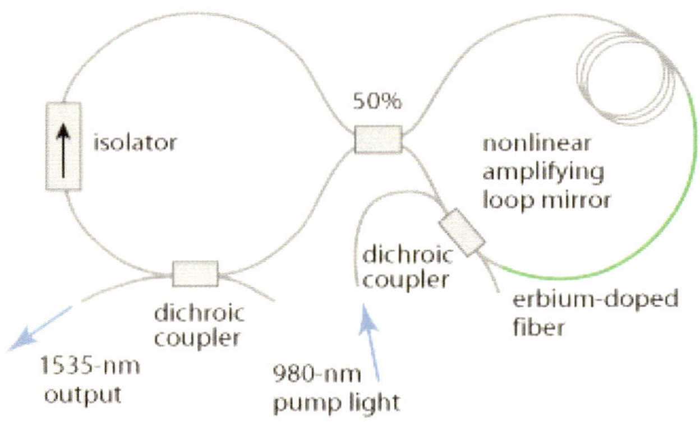

Figure-of-eight laser setup

More sophisticated resonator setups are used particularly for mode-locked fiber lasers (*ultrafast fiber lasers*), generating picosecond or femtosecond pulses. Here, the laser resonator may contain an active modulator or some kind of saturable absorber. An artificial saturable absorber can be constructed using the effect of nonlinear polarization rotation, or a nonlinear fiber loop mirror. A nonlinear loop mirror is used e.g. in a "figure-of-eight laser", as shown in figure above where there is a main resonator on the left-hand side and a nonlinear fiber loop, which does the amplification, shaping and stabilization of a circulating ultras hort pulse. Particularly for harmonic mode locking, additional means may be used, such as subcavities acting as optical filters.

Raman Fiber Lasers

A special type of fiber lasers are fiber Raman lasers, relying on Raman gain associated with the fiber nonlinearity. Such lasers usually use relatively long fibers, sometimes of a type with increased nonlinearity, and typical pump powers of the order of 1 W. With several nested pairs of fiber Bragg gratings, the Raman conversion can be done in several steps, bridging hundreds of nanometers between the pump and output wavelength. Raman fiber lasers can e.g. be pumped in the 1-μm region and generate 1.4-μm light as required for pumping 1.5-μm erbium-doped fiber amplifiers.

Fiber Lasers with Semiconductor Optical Amplifiers

There are some lasers which have a semiconductor optical amplifier (SOA) as the gain medium in a resonator made of fibers. Even though the actual laser process does not occur in a fiber, such fibers are sometimes called fiber lasers. They typically emit relatively small optical powers of a few milliwatts or even less. Sometimes they exploit the very different properties of the semiconductor gain medium, as compared with a rare-earth-doped fiber, in particular the much smaller saturation energy and upper-state lifetime. Rather than only generating coherent light, such lasers can be used for information processing in optical fiber communications systems – for example the wavelength conversion of data channels based on cross-saturation effects.

Special Attractions of Fibers as Laser Gain Media

- As fibers can be coiled and the light propagating in fibers is well shielded from the environment (e.g. concerning dust), fiber lasers can have a compact and rugged setup, provided that the whole laser resonator is built only with fiber components (*all-fiber setup*) such as fiber Bragg gratings and fiber couplers (i.e., avoiding free-space optics and any requirement for alignment).
- Fiber gain media have a large gain bandwidth due to strongly broadened laser transitions in glasses, permitting wide wavelength tuning ranges and/or the generation of ultra short pulses. Also, fiber lasers have broad spectral regions with good pump absorption, making the exact pump wavelength uncritical, so that temperature stabilization of the pump diodes is usually not necessary.
- Diffraction-limited beam quality is easily obtained when single-mode fibers are used, and sometimes also with slightly multimode fibers.
- Due to the high gain efficiency of doped fibers, fiber lasers have the potential to operate with very small pump powers. Also, it is possible to obtain very high power efficiencies.

- In recent years, the potential for very high output powers (several kilowatts with double-clad fibers) has been convincingly demonstrated.
- Again due to the guidance, which allows high pump intensities to be applied over long lengths, fiber lasers can be operated even on very "difficult" laser transitions (e.g. of upconversion lasers).

On the other hand, fiber lasers can suffer from various problems:

- When the pump light has to be launched from free space into a single-mode core, the alignment is critical. This problem can be eliminated by using fiber-coupled pump diodes.
- Most fibers exhibit a complicated temperature-dependent polarization evolution, unless polarization-maintaining fibers or Faraday rotators are used. Such measures, however, are normally not compatible with nonlinear polarization rotation mode locking.
- Nonlinear effects often limit the performance, e.g. in terms of powers achievable in single-frequency operation or the pulse quality of mode-locked lasers. For example, Kelly sidebands are often seen, whereas mode-locked bulk lasers rarely exhibit this effect.
- At high powers, there is a risk of fiber damage even below the actual damage threshold of the material (→ *fiber fuse*).
- Fibers have a limited gain and pump absorption per unit length, making it difficult to realize very short resonators e.g. for single-frequency lasers or for multi-gigahertz mode-locked lasers. However, significant progress has been made in this direction recently via the development of very highly doped fibers, usually made from phosphate glasses.

Also note that fiber lasers are in many cases substantially more difficult to design than bulk lasers. This results from very different reasons, including strong saturation effects caused by the high optical intensities, the quasi-three-level behavior of nearly all fiber laser transitions, and the complicated pulse formation mechanisms in mode-locked fiber lasers. As a result, the laser development project can be more costly.

The article on fiber lasers versus bulk lasers compares the strengths and weaknesses of fiber and bulk lasers.

Fiber Laser Modeling

Many technical aspects in fiber lasers are significantly more complicated than in bulk lasers. Reasons for that are manifold:

- Fiber lasers are typically operated with a higher gain and higher resonator losses.

- Optical intensities in fiber lasers are often far above the saturation intensity, leading to strong saturation effects (even for pump waves).
- Most active fibers have quasi-three-level gain systems, and their operation characteristics are more complicated than those of four-level lasers.
- Fiber laser systems are often more complex, for example using master oscillator power amplifier architectures.

For these reasons, the operation details of a fiber laser (or fiber laser system) can often not be understood only based on simple analytical calculations. Numerical simulations, carried out with some kind of fiber simulation software, are therefore required for calculating the possible laser performance, analyzing detrimental effects, and optimizing prototype and product designs. Such simulations can address many different technical aspects:

- Rate equation modeling can be used for calculating the behavior of single laser-active ions, or combinations of ions involving energy transfer processes.
- A mode solver, i.e., a calculator for fiber modes, can produce inputs for further calculations – in particular, mode intensity profiles.
- In some situations, numerical beam propagation is of interest. For example, this is often the case for highly multimode fibers, including the pump claddings of double-clad fibers.
- Refined algorithms are required for calculating the steady state of lasers and amplifiers, with a self-consistent solution for optical intensities and excitation densities of laser-active ions throughout the fiber. (Note that optical intensities and excitation densities mutually influence each other.)
- Dynamical models are used for calculating pulse amplification and Q switching, for example.
- Ultra short pulse propagation in fibers can also be numerically simulated under the influence of effects like laser gain, the limited gain bandwidth, chromatic dispersion, various nonlinearities, etc.

As an example for surprising features even of simple fiber lasers, Figure below shows the optical powers and excitation densities along the fiber of an Yb-doped single-mode fiber laser. A fiber Bragg grating with 25% peak reflectivity at 1030 nm on the right side serves as the output coupler, whereas a highly reflecting Bragg grating is used on the left side. The pump light (at 975 nm) is coupled in through that grating. A nearly linear (rather than exponential) decay of pump power on the left side results from strong pump saturation. The fiber is somewhat over-long, resulting in slight signal reabsorption on the right side. That reabsorption maintains a significant ytterbium excitation despite the vanishing pump power, but causes only a negligible reduction in signal output power. Losses via ASE (not shown here) are also negligible.

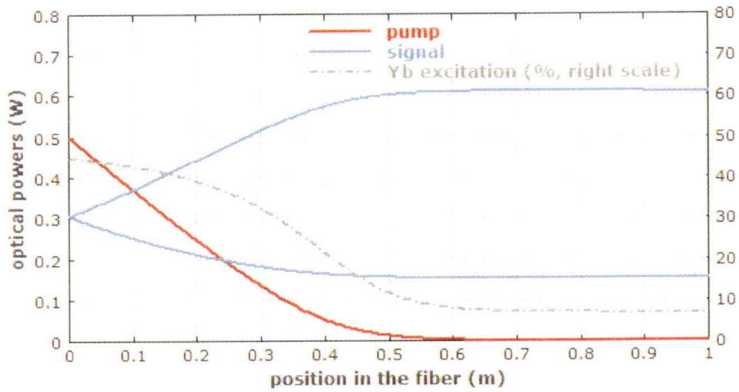

Figure: Optical powers and excitation densities along the fiber of an Yb-doped single-mode fiber laser, core-pumped at 975 nm. Note that the intracavity signal power can be higher than the pump power; only part of that power can be coupled out. The simulation has been done with the software RP Fiber Power

Figure below shows the same for a modified output coupler grating, so that lasing occurs at 1080 nm. The lower emission cross-sections at 1080 nm lead to a higher degree of Yb excitation and thus to weaker pump absorption. This demonstrates that the required fiber length depends not only on the absorption characteristics at the pump wavelength, but also on the details for the signal, such as the signal wavelength and the resonator losses.

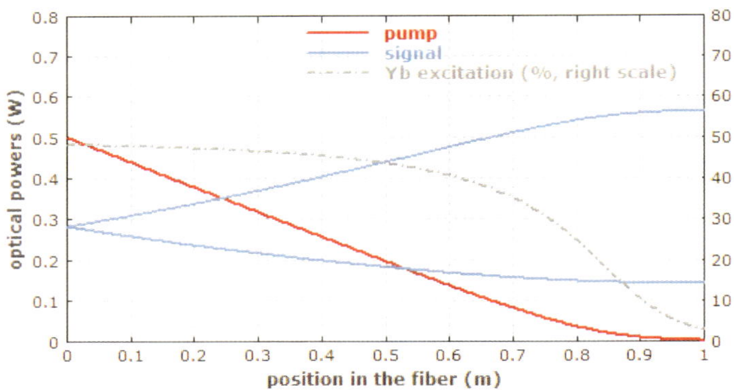

Figure: Same as in figure above, but for a fiber bragg grating for lasing at 1080 nm

If the fiber length in the last case would be reduced to 0.7 m, one might expect a moderate reduction in output power due to incomplete pump absorption. However, a simulation (not shown here) tells that lasing would stop completely, and 94% of the pump power would leave the fiber on the right side. The Yb excitation density of about 50% throughout the fiber would not be sufficient to reach the laser threshold. For a reduced pump wavelength of 940 nm, however, lasing would be possible again – despite the reduced pump absorption cross section, because pump saturation effects would be weaker.

Uses of Fibre Lasers

The first reason a fibre laser is useful is because it is stable. When we want to deliver a laser beam, we usually need an optical fibre to move it around safely. The alternative is to bounce it between mirrors, but these need to be very finely aligned. Coupling a laser beam into the tiny core of a fibre is very hard, and also needs to be finely adjusted. So normal lasers are very sensitive to being knocked or banged, and once the optics go out of alignment it takes a specialist to sort them out.

In contrast, the fibre laser generates the beam inside the fibre, so the delivery of the beam doesn't require any complicated or sensitive optics. This makes it very stable and easy to use.

The second reason a fibre laser is useful is because the beam quality is so high. The beam is generated and confined inside the small core of the fibre. In practice, this means that the beam is very straight, and that it can be focused on a very small dot. The smaller the dot, the more effective the laser will be for cutting, for example, but in many uses for lasers this quality is required.

The third reason is because they can be made to go up to very high powers. Lasers start to get to the point where they can harm your skin at around 1 Watt of power. This isn't very much power compared to a kettle or a toaster, but when it is all focused on a single point it seems like a lot! It's the difference between being gently warmed by the sun, or having a big lense focus the sun onto you, creating a powerful beam.

Fibre lasers are continually getting more powerful, and have been made with over 1,000 Watts of power (1kW). That is more than enough to cut through an inch of steel!

The fourth reason is that they are very efficient and easy to cool. Other lasers may only be able to convert a small percentage of the power put into them, but a fibre laser converts around 70-80% of the power put in by the pump source. This is better for power consumption, but it means less power is being converted to heat. The fibres can be quite long, and the heat is distributed over the length of the fibre, which protects the fibre from getting so hot that it breaks. Almost all materials change their optical properties as they heat up, so that can affect the beam quality.

As the capabilities and practicalities of fibre lasers improve, so does the range of uses they are put to. Fibre lasers are being increasingly used in manufacturing, for example.

References

- Solid-state-lasers, microwave-radar, electronics: daenotes.com, Retrieved 19 July 2018
- Eryag-laser, medicine-and-dentistry: sciencedirect.com, Retrieved 10 April 2018
- Construction-and-working-of-ruby-laser: winnerscience.com, Retrieved 29 May 2018
- Ruby-lazer, microwave-radar, electronics: daenotes.com, Retrieved 30 March 2018
- How-fibre-lasers-work: orc.soton.ac.uk, Retrieved 19 April 2018

4

Gas Lasers

A gas laser is a laser, which emits coherent light produced by the discharge of electric current through a gas. Many different gases can be used in gas lasers. This chapter closely examines the varied types of gas lasers such as helium–neon laser, carbon dioxide laser, argon ion laser, krypton ion laser, etc. for an extensive understanding of the subject.

In these lasers the lasing medium is made-up of one or a mixture of gases or vapors. Gas lasers can be classified in terms of the type of transitions that lead to their operation: atomic or molecular. The most common of all gas lasers is the helium-neon (He-Ne) laser. The presence of two atomic species (helium and neon) in this gas laser might suggest that the medium is made of molecules, but these two species of atoms do not form a stable molecule. In fact, all inert atoms like helium, argon, krypton, etc. (those in the last column of the Periodic Table) hold tightly to their own electron clouds and seldom form a molecule or react with other atoms (hence the name: inert). In the He-Ne laser the transition that produces the output light is an atomic transition. Gas lasers that employ molecular gas or vapor for their lasing medium use molecular transitions for their lasing operation. Molecular transitions tend to be more complex than atomic ones. As a consequence the laser light produced by molecular lasers tends to have a wider and more varied collection of properties. Examples of some common molecular gas lasers are carbon monoxide (CO), carbon dioxide (CO_2), excimer, and nitrogen (N_2) lasers.

A variety of lasers is based on gases as gain media. The laser-active entities are either single atoms or molecules, and are often used in a mixture with other substances having auxiliary functions. A population inversion as the prerequisite for gain via stimulated emission is in most cases achieved by pumping the gas with an electric discharge, but there are also gas lasers using a chemical reaction, optically pumped devices, and Raman lasers. During operation, the gas is often in the state of plasma, containing a significant concentration of electrically charged particles.

Most gas lasers emit with a high beam quality, often close to diffraction-limited, since the gas introduces only weak optical distortions, despite considerable temperature gradients. Their operation usually requires a high-voltage supply, often with a high electrical power. Some high-power gas lasers use a system for quickly circulating the gas (*forced convection, fast flow*).

Types of Gas Lasers

There are very different kinds of gas lasers, operating in entirely different regimes concerning emission wavelength and output power:

- *Helium–neon lasers* (He–Ne lasers): often emit red light at 632.8 nm, but can also be made for other wavelengths such as 543.5 nm (green), 594.1 nm (yellow), 611.9 nm (orange), 3.39 μm, or 1.15 μm. Typical He–Ne lasers have a gas cell with a length of the order of 20 cm and generate a few milliwatts of output power in continuous-wave operation at 632.8 nm, using several watts of electrical power. Helium–neon lasers are often used for alignment and in interferometers, and compete with laser diodes, which are more compact and efficient. Some HeNe lasers serve in optical frequency standards.
- *Argon ion lasers:* use a typically larger (e.g. 1 m long) water-cooled tube with an argon plasma, made with an electrical discharge with high current density in order to achieve a high degree of ionization. They can generate more than 20 W of output power in green light at 514.5 nm, and less at some other wavelengths such as 457.9, 488.0, or 351 nm. Their power efficiency is fairly low, so that tens of kilowatts of electrical power are required for multi-watt green output, and the cooling system has corresponding dimensions. There are smaller tubes for air-cooled argon lasers, requiring hundreds of watts for generating some tens of milliwatts. Argon ion lasers can be used e.g. for pumping titanium–sapphire lasers and dye lasers, and are rivaled by frequency-doubled diode-pumped solid-state lasers.
- *Krypton ion lasers:* are similar to argon ion lasers and can emit high powers at 647.1 nm and some other wavelengths.
- *Carbon dioxide lasers* (CO_2 lasers): use a gas mixture of CO_2, helium (He), nitrogen (N_2), and possibly some hydrogen (H_2), water vapor, and/or xenon (Xe) for generating laser radiation mostly at 10.6 μm. They have wall-plug efficiencies above 10% and are suitable for output powers of multiple kilowatts with fairly high beam quality. They are widely used for material processing, e.g. cutting, welding and marking, but also in laser surgery. CO_2 lasers are called molecular lasers because laser-active molecules are used.
- *Carbon monoxide lasers* (CO lasers: can have wall-plug efficiencies of the order of 40%, thus being substantially more power-efficient than CO_2 lasers. They can emit on various lines between 4.8 μm and 8.3 μm and are mostly used as light sources for laser spectroscopy. Following technological advances concerning the device lifetime, CO lasers emitting around 5.5 μm might also become interesting for laser material processing (e.g. cutting of glasses); in comparison with the widespread CO_2 lasers, they offer better absorption in many materials and better focusing capabilities.

- *Excimer lasers* (rare gas halide lasers, exciplex lasers): are also pumped with an electrical discharge, but in that case the pumping energy is used to form unstable molecules which can emit photons when disassociating. Most excimer lasers are ultraviolet lasers and are operated with current pulses, leading to the emission of intense nanosecond pulses. They are used for various types of material processing, including pulsed laser deposition, laser marking, and the fabrication of fiber Bragg gratings. There are also medical applications e.g. in ophthalmology.
- *Nitrogen lasers:* are another type of pulsed ultraviolet laser, based on pure nitrogen, a nitrogen–helium mixture, and sometimes even simply air (with lower performance). Emission typically occurs at 337.1 nm in the form of short pulses; a self-terminating laser transition is used. The high gain leads to relatively efficient super luminescent emission even without a laser resonator. Nitrogen lasers are relatively easy to build and operate, and have been made by many hobbyists without refined laboratory equipment.
- *Hydrogen lasers:* can be used to access even shorter wavelengths around 160 nm, 123 nm or 116 nm.
- Various *metal vapor lasers:* use a metal vapor, excited and heated by an electric discharge.

Copper vapor lasers: are excited with intense current pulses and generate nanosecond pulses at 510.6 nm (green) or 578.2 nm (yellow). The average output power can exceed 100 W. *Helium–cadmium lasers* are more similar to helium–neon lasers, emitting continuously at 442 nm (blue) or 325 nm (ultraviolet), with optical powers of the order of 100 mW. The laser transition occurs in Cd^+ ions, which become excited in collisions with excited helium atoms. - There are *alkali vapor lasers*, using e.g. a cesium or rubidium cell in an oven as the gain medium. Such lasers can be pumped with laser diodes. The power efficiency can be fairly high; note that the quantum defect can be small for a typical pumping scheme where one pumps from the ground state $^6S_{1/2}$ to $^6P_{3/2}$ and uses a transfer via a buffer gas (e.g. ethane) to the nearby $^6P_{1/2}$ as the upper laser level. At the same time, the beam quality can be much higher than that of the pump source, so that such a laser acts as an efficient brightness converter. - *Chemical lasers* convert chemical energy of gases into laser light (typically in the mid- or near-infrared region) with powers up to the megawatt level. There are e.g. hydrogen-fluoride (HF) lasers, fueled with H_2 and F_2, which is converted to HF, and oxygen-iodine lasers (COIL). Chemical lasers are mainly studied for military purposes, e.g. as anti-missile weapons, to be operated even on board large airplanes. - *Raman gas lasers* are Raman lasers, based on optical amplification via stimulated Raman scattering rather than on stimulated emission of excited ions. They can e.g. use a hydrogen cell, and need to be optically pumped.

Gas lasers can also be grouped according to the nature of their laser-active species:

- *Neutral atom gas lasers* include helium–neon lasers and copper vapor lasers.
- *Ion lasers* use free ions; examples are helium–cadmium lasers, argon ion lasers and krypton lasers. Typically, ion lasers generate shorter wavelengths, but with moderate power efficiency.
- *Molecular gas lasers* use gas molecules; Examples are carbon dioxide and carbon monoxide lasers, nitrogen lasers, and excimer lasers.

Many gas lasers have self-terminating laser transitions, where the lower state has a long lifetime. Lasing stops once the lower-state population becomes too high. Examples for such gas lasers are nitrogen lasers and copper vapor lasers. Excimer lasers can also only be operated in pulsed mode, although for different reasons.

Application of Gas Lasers

A primary reason for using certain gas lasers instead of solid-state lasers is that they offer special wavelengths, which are otherwise difficult to obtain. Another interesting aspect is that relatively high optical powers can be obtained with gas lasers; compared to solid-state lasers, particularly diode-pumped ones, the price depends less strongly on the required output power level. A good example for both aspects is the carbon dioxide laser, which is quite a unique long-wavelength source with high output power. Similarly, excimer lasers provide high powers in the ultraviolet spectral region.

The helium–neon laser has been widely used for the generation of red laser light, but is now increasingly replaced by cheaper and more compact laser diodes. Similarly, argon ion lasers were often used for pumping titanium-sapphire lasers, but are now often replaced by frequency-doubled solid-state lasers.

Helium–Neon Laser

He-Ne stands for Helium-Neon. He-Ne laser active medium consists of two gases which do not interact form a molecule. Therefore He-Ne laser is one type of atomic gas lasers.

Construction of He-Ne Lasers

The construction of typical He-Ne laser plasma tube can be shown as:

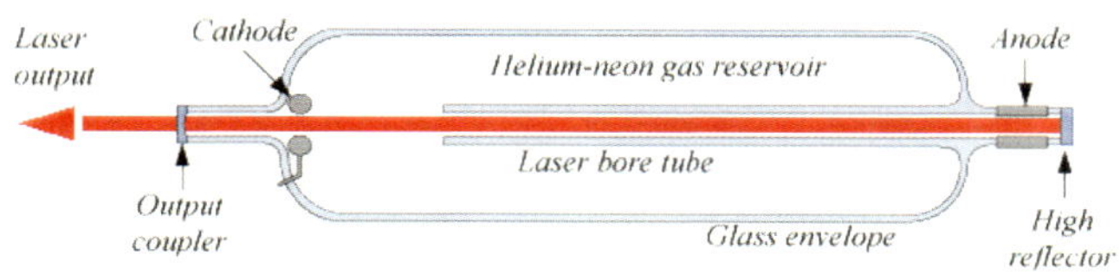

The tube where the lasing action takes place consists of a glass envelop with a narrow capillary tube through the center. The capillary tube is designed to direct the electrical discharge through its small bore to produce very high current densities in the gas.

The output coupler and the HR (high reflective) mirror are located at the opposite ends of the plasma tube. To make laser tubes more economical and durable manufacturers often attach the mirror s directly to the ends of the capillary tube as shown above. This is very common with small low power lasers. With high power tubes or when optically polarized output is desired, the capillary tubes ends are cut at an angle and sealed with glass planes called Brewster windows. When this is done then the mirrors mush be mounted in mechanically stable but adjustable mounts. This allows the operator to align the mirror surfaces parallel to each other but perpendicular to the axis of the capillary tube.

The plasma tube has a large cylindrical metallic cathode and a smaller metallic anode.

The current is directed from cathode to anode. In figure shown, the gas reservoir provides a supply of extra gas. This reservoir helps to maintain a uniform pressure over long period of time and provides extra gas to replace any gas that may escape through the tube or through the seals where the loads pass through the glass envelop. Usually all He-Ne plasma tubes have a gas reservoir.

Function of He-Ne Laser

In the He-Ne laser the light is produced by atomic transitions within the Neon atom. The Helium does not directly produce laser light but it acts as a buffer gas, the purpose of which is to assist/help the atoms of the other gas to produce lasing in as manner.

When energy from the pumping source is applied He-Ne gas mixture then some of the energy is observed by the Helium atoms. In other words we can say that helium atoms achieve an excited state. Now when the Helium atoms move within the laser tube, they collide with the Neon atoms. At each collision some of the energy within the helium atom is transferred to the Neon atom and so raising it to an excited meta-stable state. When a sufficient number of Neon atoms reach to this state then population inversion occurs and hence the lasing can take place.

This can be shown by simplified energy level diagram as:

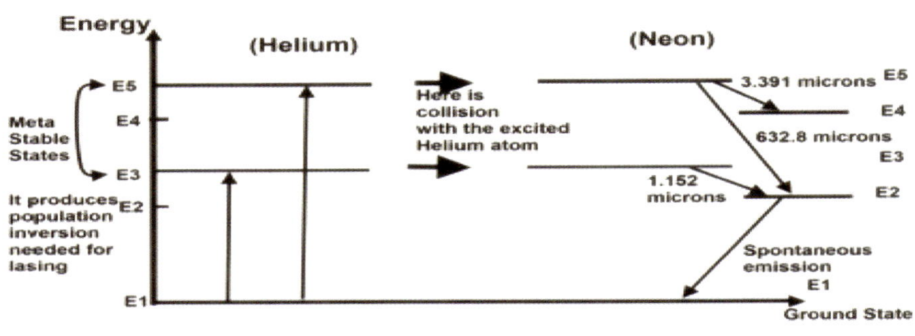

Here upward transition shows the absorption of energy from the pumping source by Helium atom. While down ward transition shows the emission of energy / light or lasing present in the Neon atom only.

In figure above there are 3 down word energy transitions for Neon that produce lasing. If transition occurs at the relatively small energy step from E5 to E4 then low energy infrared photon is released with a wavelength of 3.391 microns. If transition occurs at E5 to E2 which is much larger energy step then it produces short wavelength more energetic photon at 632.8nm. This gives the red light which is most desirable for He-Ne laser applications.

E3 to E2 then it produces a laser output at 1.152microns in infrared portion of the spectrum.

In all He-Ne lasers the feedback mechanism consists of pair of coated mirrors. The coating is usually reflecting mirror and 95%-99% of the light at output coupler. The reflection at the output mirror must be higher if the active medium is short in length because the gain of the active medium is low. If the active medium is longer more gain is produced an a larger percentage of the beam can be provided as an output. Therefore in He-Ne laser with a longer active medium, the reflectivity of the output coupler can be less.

Characteristics of He-Ne Laser

The He-Ne laser is a relatively low power device with an output in the visible red portion of the spectrum. The most common wavelength produced by He-Ne lasers is 632.8nm, although two lower power (1.152μm and 3.391μm) infrared wavelengths can be produced if desired. Majority of He-Ne lasers generate less than 10m watt of power, but some can be obtained commercially with up to 50m watts of power. For He-Ne lasers the typical laser tube is from 10 to 100 cm in length and the life time of such a tube can be as high as 20,000 hours.

Applications / Uses of He-Ne Laser

The Helium-Neon gas laser is one of the most commonly used laser today because of the following applications.

- He-Ne lasers are produced in large quantities from many years.
- Many schools / colleges / universities use this type of laser in their science programs and experiments.
- He-Ne lasers also used in super market checkout counters to read bar codes and QR codes.
- The He-Ne lasers also used by newspapers for reproducing transmitted photographs.

- He-Ne lasers can be use as an alignment tool.
- It is also used in Guns for targeting.

Advantages of He-Ne Laser

- He-Ne laser has very good coherence property.
- He-Ne laser can produce three wavelengths that are 1.152μm, 3.391 μm and 632.8nm, in which the 632.8nm is most common because it is visible usually in red color.
- He-Ne laser tube has very small length approximately from 10 to 100cm and best life time of 20.000 hours.
- Cost of He-Ne laser is less from most of other lasers.
- Construction of He-Ne laser is also not very complex.
- He-Ne laser provide inherent safety due to low power output.

Disadvantages of He-Ne Laser

The weak points of He-Ne laser are:

- It is relatively low power device means its output power is low.
- He-Ne laser is low gain system/ device.
- To obtain single wavelength laser light, the other two wavelengths of laser need suppression, which is done by many techniques and devices. So it requires extra technical skill and increases the cast also.
- High voltage requirement can be considered its disadvantage.
- Escaping of gas from laser plasma tube is also its disadvantage.

Suppressing Unwanted Wavelengths in He-Ne Laser

The He-Ne laser produces three different wavelengths that are 1.152μm, 3.391 μm and 632.8nm. At the 1st two the infrared photons are released while at the 3rd one i.e. 632.8nm more energetic photon of red color are released because in spectrum at 632.8 approximately red color occurs.

Since for He-Ne laser most applications, the red output is most desirable therefore it is necessary to suppress the infrared wave lengths and prevent them from lasing so that the energy directs into the desired output. In small He-Ne laser, it is normally done with proper coatings on the feedback mirrors. If they are coated for high reflectance at 632.8nm then they will not reflect a high percentage of eh infrared wavelengths.

This causes the desired wavelength to have a low loss and the undesired wavelength to have a high loss, which reduces the possibility in the infrared wavelengths lasing and enhancer lasing at 632.8nm.

For high power He-Ne lasers the mirror coating cannot totally suppress the infrared wave length. Therefore in these lasers the magnet is placed near the plasma tube which provides an additional suppression effect on the infrared wavelength.

If these two suppression techniques are not enough to prevent infrared lasing than an infrared filter can be used between the laser mirrors to absorb the unwanted wavelengths.

Carbon Dioxide Laser

C.K.N.Patel invented this molecular laser in 1964, at Bell Telephone Labs, USA. The active medium responsible for lasing is the carbon dioxide molecules, giving output at 10.6 μm and 9.6 μm. The very first CO_2 laser produced only a few milliwatt output. CO_2 lasers typically emit at a wavelength of 10.6 μm, but there are other lines in the region of 9-11 μm (particularly at 9.6 μm). In most cases, average powers are between some tens of watts and many kilowatts. The power conversion efficiency can be between 10 - 20 %. It is higher than for most flash lamp pumped solid-state lasers, but much lower than for diode pumped solid-state lasers. With the technological advancements, the present class of CO_2 lasers produces CW output of megawatts.

The salient features of fundamentals of lasing action in CO_2 laser are as follows. Carbon dioxide molecule is a tri-atomic molecule consisting of two oxygen atoms covalently bonded to a central carbon atom. It has three fundamental modes of vibration, namely, symmetric, bending and asymmetric stretching modes, which are shown in figure below.

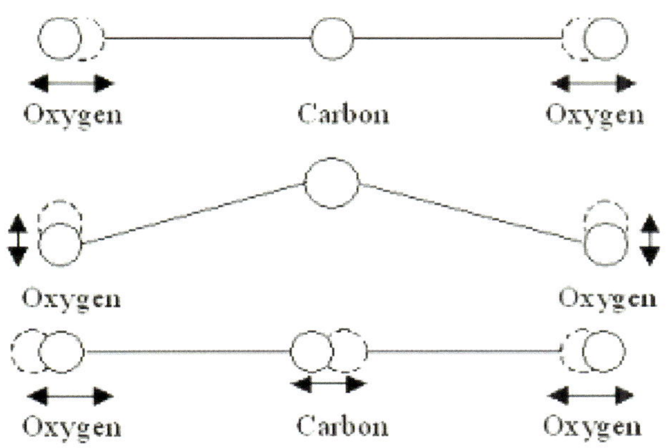

Modes of Vibration of CO_2 molecules

In the symmetric mode, carbon atom is in the center and the two oxygen atoms oscillate symmetrically along the axis of the molecule in unison, either away from or towards each other. In bending stretch mode, the oscillation of the molecules is in perpendicular direction to the axis. In the asymmetric mode, though the molecules oscillate along the axis, only one of the oxygen atoms comes close to the central carbon atom at a time and as this atom moves away from the center, the other atom comes towards the carbon atom and they alternate the movements.

The three vibrational states of CO_2 are referred to by the number of vibrational quanta n1, n2 and n3 related to the symmetric, bend and asymmetric stretch modes respectively. It may be noted that it is possible for a molecule to execute all the three modes at a time or a molecule excited to a level can have more than one quantum of energy in any one of the modes or in all of the three modes. For example, (001) means that the CO_2 molecule has a single vibrational quantum in asymmetric stretch mode.

There are many vibrational energy exchange processes that take place between different molecules in the mixture of gases. Transitions between vibrational energy levels results in emission in the infrared, where as transitions between rotational states emit photons in the microwave region. Gas mixture in the CO_2 laser consists of helium, nitrogen and carbon dioxide gases. A total pressure of 6-20 torr is made up of 10 to 20% N_2 and 10 to 15% CO_2 and rest being helium. Important vibrational relaxation processes that occur in CO_2 and N_2system are shown in the figure given here.

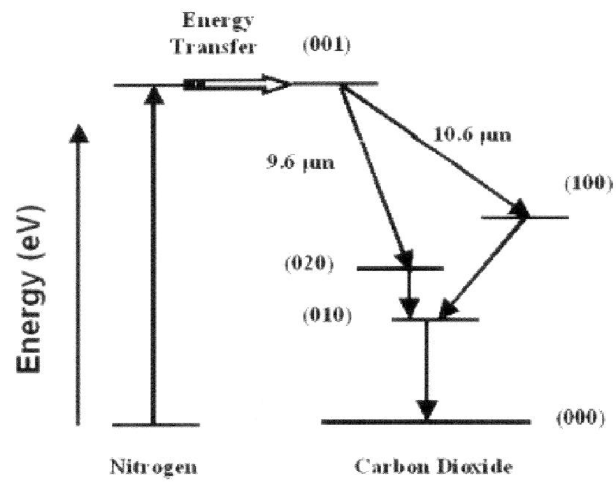

Simplified exchange between energy levels in $CO_2 - N_2$ system

The laser transition is between upper laser level (001) and the lower laser level (100), emitting out put at 10.6μm. Lot of transitions takes place for generating 10.6μm laser emission. By electron impact nitrogen (N_2) excitation takes place to vibration level, ν = 1. Transfer of energy between nitrogen and the nearly resonant first asymmetric stretch level (001) of carbon dioxide molecules takes place, since the restoring force constant of N_2 and that of CO_2 molecules are almost identical and results in populating the upper laser level. It may be noted that the population is shared between n1 and n2 modes and the vibrational energy in the n2 manifolds is converted in to translational energy by collisions with helium. (100) state and (020) state have only even spin members. Further, members of the same vibrational states are in thermal equilibrium and the available energy is redistributed between them, as determined by Boltzmann statistics. Two important functions are carried out by helium, namely, it maintains plasma discharge and also helps in depopulating the lower laser level. Collision between carbon dioxide and helium atoms results in the transfer of energy to the helium atom.

After the completion of laser transition from (001) state to (100) state, CO_2 molecules still have lot of energy, which they have to loose before getting excited back to the higher level. This takes place in two steps. i.e. first from (100) to (010) and then from (010) to (000) level, which is the ground state.

It would not be out of place to discuss the role of nitrogen and helium in the lasing of CO_2 laser. Nitrogen has only one vibrational mode as it is a diatomic molecule and it has only one vibrational quantum number. With reference to the figure above, it can be seen that nitrogen level ($v = 1$) is very near to the CO_2 (001) level. Collision between CO_2 molecule in the ground state (000) and N_2 molecule ($v = 1$) results in the transfer of energy to CO_2 molecule. Consequently, CO_2 molecule will be at (001) state and N_2 will be at ($v = 0$) state. In an electric discharge laser, nitrogen is used to transform the available energy to excite CO_2 molecules to the upper lasing level.

De-excitation of CO_2 from (100) level to (010) level is very efficient. But the relaxation from (010) level to (000) state is very inefficient in pure CO_2. This is due to the fact that the de-excitation is by collision and the rate depends on the nature of the particles involved. At a pressure of say 1 torr, each CO_2 molecule under goes about 100 collisions per second, where as in the presence of helium gas, the rate increases to about 4,000, due to which the de-excitation is accelerated greatly, thus increasing the efficiency of the system. Water vapour or helium is used as a catalyst to improve the efficiency of the CO_2 laser. Though the collision rate in the presence of water vapor is about 100,000 per second at a pressure of 1 torr, it helps only in depopulating the lower laser level, where as helium helps in de-excitation of the lower laser level as well as in maintaining the excitation to upper level.

At this juncture, we would like to draw the attention of the reader to certain specific factors related to the design of CO_2 lasers, namely methods of excitation and cooling of laser gas mixtures. Electrical discharge (radio frequency and direct current), gas dynamic (thermal) and electron beam (high energy) excitation techniques have been employed to generate high power output from CO_2 lasers. Like other lasers, CO_2 laser is also not a very efficient device. It has efficiency about 20%, the rest of the energy going as heat and as such cooling is of great importance in the design and development of CO_2 lasers. Unused energy increases the temperature of the system and the lower level population cannot be emptied fast enough to ground state. This naturally affects the population inversion and laser action ceases altogether due to temperature rise.

CO_2 laser uses CO_2, N_2, He and sometimes some hydrogen (H_2) and or water vapor mixtures. The role of hydrogen or water vapor (2-5 %) is to help (particularly in sealed-tube lasers) to reoxidize carbon monoxide (formed in the discharge) to carbon dioxide. It may be mentioned that not only the total pressure of laser gas, but also the proportion at which they are mixed are also very important to produce maximum output from the system. For example, to start with one may begin with 10-20 % CO_2, 10-20% N_2 and the rest 60 - 80 % being He. Then the proportion of these gases may be optimized to get maximum output. Here one cannot forget the importance of current, higher the current, higher is the laser output. Such a laser is electrically pumped via a gas discharge, which can be operated with DC current, with AC current (e.g. 20-50 kHz) or in the radio frequency (RF) domain.

The gas mixture in a carbon dioxide laser is subjected to an electric discharge causing the low-pressure gas to form plasma. In the plasma, the molecules take up various excited states as expected from the Boltzmann distribution. Some will be in the upper state (001), which represents an asymmetric oscillation mode. This molecule may lose its energy by collision with the walls of the cavity or by spontaneous emission. Through spontaneous emission the state falls to the symmetric oscillation mode (100) and a photon of light of wavelength 10.6 μm is emitted traveling in any direction. One of these photons, by chance, will be traveling down the optics axis of the cavity and will start oscillating between the resonator mirrors.

CO_2 laser can work in the pulsed as well as in the CW mode. Not only DC excitation generate CW output, but high frequency (HF) and radio frequency (RF) can also generate CW output, since the life times of the excited molecules are long when compared with HF and RF frequency. Maximum pulse frequency of the system is limited by the speed of response of the discharge.

Basically, CO_2 laser system arrangement is very similar to other gas lasers, i.e. a gas filled tube with a pair of mirrors at the ends and the gas excited with DC or RF electrical discharge. Total reflecting mirror is a highly polished solid molybdnium or silicon with high reflectivity coatings or gold-coated copper. Output coupler is normally Zinc Selenide (ZnSe) as it has very low transmission losses at CO_2laser wavelength. Germanium is another choice, but this has to be cooled, especially for high output. For high power applications, gold mirrors and zinc selenide windows and lenses are preferred. Recently diamond windows and even lenses are also being used. Diamond windows are extremely expensive, but their high thermal conductivity and hardness make them useful in high-power applications.

The most basic form of a CO_2 laser consists of a gas discharge (with a mix close to that specified above) with a total reflector at one end, and an output coupler (usually a semi-reflective coated zinc selenide mirror) at the output end. The reflectivity of the output coupler is typically around 5-15%. The laser output may also be edge-coupled in higher power systems to reduce optical heating problems.

CO_2 lasers have been operated in the following modes:

- Sealed or no-flow
- Slow axial flow
- Fast flow
- Transverse excited atmosphere (TEA)
- Wave guide type
- Gas dynamic mode.

The maximum attainable power output depends upon the efficiency of excitation and the cooling of the Laser gas. Thus the practical designs of the Laser vary according to the form of selective excitation, the cooling of the gas and with it the attainable power output. The development of a cost effective power Semiconductor for the high frequency region made it possible to realise high frequency excited discharging. They also have the advantage against parallel flow discharge as the power can be connected to the Laser gas almost without losses. The way the waste heat is rejected (or gas cooling method) has a large influence on the laser system design. In principle, it can be performed by two possible methods. The first method is based on the conventional process of natural diffusion of the heated gas to the tube wall, which is the operating principle of the sealed, and slow axial flow lasers. The second method is based on the gas forced convection, which is the operating principle of the fast flow lasers.

Sealed off Laser

A sealed off laser is a laser in which the ends of the gain tube are closed off and the gas is not allowed to flow into or out of the tube. The scheme of the Laser systems as a whole, at first seems rather simple; but the various details adopted for system life are quite complex. Contamination of the laser gas due to sputtering of the electrodes and production of heat are the major problems encountered in the working of the system. The electrodes must be made of a precious metal so that the continual material erosion does not react with the gases and especially with its disassociation products. In this case, heat is removed by helium removing the excess energy and moving to the walls of the gain tube where it gives up its energy to the wall through collisions. Another issue that arises in this case is that carbon monoxide is formed, which will cause laser action to cease. The addition of water vapour takes care that the formed CO gas will react with it and generates CO_2 and finally re-mixes again with the CO_2 gas. These types of lasers are capable of hundreds of watts of output power. Optimised systems achieve power outputs of up to 60 - 70 Watt per meter of discharged length. The cooling of gases takes place through diffusion- that means that the gas molecules diffuse from the center of the discharge to the wall, where they are cooled by collisions.

For laser powers between a few watts and a several hundred watts, it is common to use sealed-tube or no-flow lasers, where the laser bore and gas supply are contained in a sealed tube. Such lasers are compact and rugged, and reach operation lifetimes of several thousands of hours.

Slow Flow Laser

Contamination and thermal aspects can be overcome to a great extent by flowing the laser gases, where the electrodes are outside the discharge tube. These are the systems required for initial understanding of CO_2 Laser development and are still very common keeping in view the complications associated with the sealed off lasers. For experimentation, this system is ideal with regard to Laser behaviour and basic capabilities. With an optimised system having an excitation length of 1 meter, Laser power of up to a maximum of 100 Watt can be achieved. The excitation takes place over a longitudinal direct current discharge, The excitation of the initial Laser level is achieved firstly by electron collision and secondly due to collisions with excited N_2 molecules. The important features are:

- Water or oil cooling in a double-walled glass plasma tube.
- Gas flow rates (1-20 liters per minute, depending upon size and output of laser).
- Dc excitation, coaxial with gas flow and laser beam.
- Low-current (3 - 100 mA) operation.
- Gas pressures in the range of 10 to 30 torr.
- Tube diameters of 1.0-2.0 cm.
- Available output powers up to about 50 -100 W per meter of tube length.

The primary factor that limits output power of these lasers is their inability to efficiently remove waste heat from the gas. Cooling is principally achieved by helium (He) collisions with tube walls. Air-cooling of CO_2 laser tubes is possible, but this results in an elevated wall temperature and greatly reduces laser efficiency. Smaller CO_2 lasers and those used in research often employ water-cooling. Industrial CO_2 lasers usually use recirculating oil and oil-to-water heat exchangers for better system stability and reduced maintenance. An increase of tube current beyond the recommended operating value results in more heat than can be effectively removed from the system in this manner. Increases in tube diameter also decrease cooling efficiency by increasing the path length necessary for (He) atoms to reach the walls from the center of the tube. Thus, the only effective method of increasing output power of this type of CO_2 laser is to extend the active length. For best results, this must also be accompanied by an increase in gas flow rate. In larger systems the gas is re-circulated with a few percent being replaced on each cycle.

Slow axial flow lasers can generate few hundreds of watts of output, typically about 50 to 70 watts per meter length of the tube, independent of the tube diameter. To generate hundreds of watts, few meters long tube will be required, which is not very practical. In order to overcome this problem, folded tube technique has been employed. But the folding mirrors increase the complexity of the system. The optical quality of the laser beam is also affected adversely. Further, alignment of the mirrors and cleaning of the same poses practical difficulty in maintaining the system. The laser gases are cooled by direct cooling of the gases by conduction with the walls, where the cavity walls are cooled by water jackets Typically slow flow CO_2 lasers generate a maximum of 400watts CW output with a gas flow of 20 l/minute and a coolant flow of 7 l/minute using a water jacket.

- In no flow low-power sealed-tube lasers and in slow gas flow lasers; the beam quality can be very high.
- The beam size and divergence angle are between 1 - 7 mm and 2 - 6 mrad respectively. Since the laser is being operated at relatively small pressures, the more dominant form of broadening is Doppler broadening.

Fast Flow Laser

To generate few kilowatt output, one has to go for fast axial flow CO_2 laser systems. With this system, it is possible to obtain 4-5 KW outputs, which depends on the rate of mass of flow rather than on the length of the tube, unlike in the case of slow flow system. Typical rate of gas flow is about 400 l/second and gas is recalculated after cooling through a heat exchanger. Fast axial flow lasers generally operate at less than 150mbar. Another aspect is that the modes can be controlled with the diameter of the laser tube as it acts as an optical aperture. The hot Laser gas is sucked off from the discharge chamber with the help of a Roots vacuum pump. In the process it passes through the heat exchanger. The laser gas mixture is cooled when it passes through the heat exchangers kept out side the laser cavity. Fast axial flow CO_2 laser systems are complex due to size and power requirements of re-circulation pumps and power supplies. The power output is limited to 5KW. Increasing the diameter of the laser tube increases the beam divergence, especially in the higher modes and the reduction generates thermal lensing and consequently increases beam divergence, distorting the beam. For Laser powers higher than 5 kW, however, the transversal excitation system of is more advantageous, because here, due to the far lower electrode distance, the necessary ignition and drop voltage is comparatively low. Lasers of this type are capable of producing several kilowatts of power with fast flow geometry.

Transverse Excited Atmosphere (TEA)

The pressure in the CO_2 laser tube is of the order of millibars and output is limited to 5 kilowatts. By operating the laser at a pressure of about one atmosphere as well as by passing a pulsed current transversely through the laser gas, it is possible to produce pulsed energy output of hundred of joules with pulse width ranging from microsecond to nanosecond region and pulse repetition rate ranging from single shot to 300 Hz. This type of laser is referred to as TEA (transversely excited atmospheric) laser. TEA lasers are operated in pulsed mode only, as the gas discharge would not be stable at high pressures, and are suitable for average powers of tens of kilowatts Unlike sealed off and slow flow lasers, where problem of heat disposal is solved through convection and diffusion, the development of a pulsed CO_2 Laser forms the basis of TEA laser to solve the problem of heat disposal of the Laser gas, mainly through the utilisation of the specific heat of the Laser gas. This eliminates the continuous operation, because through the increased number of collisions, inversion is destroyed. However, if the excitation pulse is faster than the destruction of inversion, then a Laser pulse can be generated whose time constant depends on the mechanisms time constant. The gas flow is low and the gas pressure is high. The excitation voltage is around tens thousand volts. The laser beam energy distribution is uniform over a relatively large area. As the voltage required for a longitudinal discharge would be too high, transverse excitation is done with a series of electrodes along the tube. Its peak power is very high up to 10^{12} W as its pulse duration is very small. However it is very difficult to focus its laser beam to a small spot due to multimode operation. Lasers of this type are used in big research plants under extremely high peak power to examine the nature of matter. A CO_2 Laser system built in the scientific laboratory in Los Alamos has a terawatt of power with a nanosecond pulse width.

Wave Guide CO_2 Laser

Wave-guide CO_2 laser is a very compact laser. Basically it has two electrodes separated by an insulator. Since the electrode separation is only a few millimeters, it acts as a bore of same dimension and the beam propagation is in the 'wave guide mode'. Normally sealed laser tube has a reservoir. High-pressure operation due to small bore allows rapid heat removal, leading to high gain and consequent high output. The beam diameter and divergence of the beam are 1 - 2 mm and 3 - 5 mrad respectively.

Gas Dynamic CO_2 Laser

There are gas dynamic CO_2 lasers for multi-megawatt powers for military applications, where the energy is not provided by a gas discharge but by a combustion process just like rocket engine. These types of lasers are a class in itself and will be discussed separately

Q - switched CO_2 Lasers

Fast pulsing has proven to be more difficult. The simplest way to pulse the laser is to switch the RF power source to excite the CO_2 plasma. Due to the high life time of the Laser output level (001) of the order of msec, a switching technique is possible for the production of higher peak power. With the correct duty cycle, one can produce pulses in the range of 100 microseconds to 10 ms.

A shorter pulse is also possible by using an internal electro-optic modulator, such as a Q-switch (or cavity dumper). It is also very easy to actively Q - switch a CO_2 laser by means of a rotating mirror or an electro-optic switch, giving rise to Q-switched peak powers up to gigawatts (GW) of peak power. Previously, Q-switching was limited to military and very high-value applications, because the limited availability of cadmium-telluride (CdTe) modulator crystals. Even then, the modulators had a short lifetime because of poor damage threshold properties of these crystals. However, recent innovations have eliminated these drawbacks. Now, a modulator with advanced growth techniques suitable for CdTe crystals produces devices with very high optical damage threshold.

Properties of some resonantly absorbing molecules have also been experimentally investigated by making use of Q-switching techniques. SF_6 has been used to passively Q-switch CO_2 lasers.

Applications

There are numerous applications for CO_2 lasers. The ranges of power that can be produced by the CO_2 laser, the relatively high efficiency, and low cost of the CO_2 laser make it a very strong candidate for the materials industry as well as the medical profession.

- The biggest use of these lasers is for material processing. CO_2 lasers are used for cutting materials such as plastic or metal, welding, etching or engraving materials. Relatively high power CO_2 lasers are frequently used in industrial applications for cutting and welding, while lower power level lasers are used for engraving. Cutting of plastic materials, wood, die boards, etc which exhibits high absorption at 10.6 μm, require moderate power levels of 20-200 W, whereas cutting and welding metals such as stainless steel, aluminum or copper, require multi-kilowatt powers.
- Recently, they have been receiving a lot of attention for use in medical procedures. They are useful in surgical procedures because water present in most of the biological tissues absorbs this frequency of light very well. CO_2 lasers have been used in surgery to cut skin, stop minor bleeding during surgery, remove or vaporize abnormalities and to perform skin resurfacing etc. Also, it could be used to treat certain skin conditions such as removal of embarrassing or annoying bumps, podules, etc.
- Because of excellent beam quality, the sealed or no flow CO_2 laser is often used in beam-deflected laser marking.
- The TEA CO_2 laser is often used in mask marking.
- They have been used as a tool to measure distance.
- Because the atmosphere is transparent for CO_2 wavelength,these lasers are also being used for military range finding and LIDAR applications.

- Further, the long operation wavelength of CO_2 lasers makes them almost eye - safe particularly at lower intensities.

Argon Ion Laser

Argon is the name of gas. The argon laser active medium consists of this gas, where the lasing takes place due to ions of argon gas. That is why argon laser is the most popular type of ion lasers.

Remember that to form ions from argon atom, electrons are removed from argon atom by applying proper high energy or voltage.

Characteristics of Argon Laser

- All ion lasers (such as argon, Krypton and mix gas) produce several different wavelengths so they are designed to produce these wavelengths either simultaneously or individually.
- The argon laser produce nine different wavelengths in which two are most powerful and most commonly used. These are in the blue and blue-green spectral regions, which are at 488nm and 514nm respectively. Argon lasers are generally operated in continuous wave mode although it is possible to mode lock them to get very rapid, short duration pulses.
- For argon lasers the output power can ranges from few mill watts up to 20 watts.
- The argon lasers have a long narrow laser cavity, which produces relatively high power TEM0,0 beam output with very low divergence typically about 1 milli radian.
- The overall efficiency of the argon laser varies from about 0.01% to 0.1%, with average efficiency about 0.02% with efficiency in these ranges, it is necessary to apply large amount of power to the laser, in order to get an output. For example if a laser has an efficiency of 0.1% then it is necessary to apply 1000 watts of power to laser in order to achieve an output of 1 watt. Remember that in these like cases the remainder of the input power is dissipated as heat.

Construction of Argon Lasers

The argon lasers are constructed differently from the He-Ne laser. Because argon lasers have efficiency very low, so needs large amount of power as input, which produces the tremendous amount of heat. A typical argon laser can be shown as:

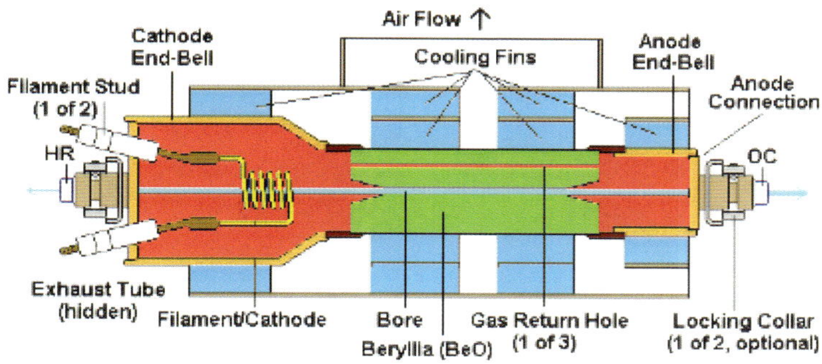

The figure shows that in argon laser there are cooling fins on the outside of the tube. This is one method of dissipating the heat generated during the lasing process. Sometime instead of cooling fins, the water jacket surrounds the tube where water is forced through the jacket to cool the tube. This water jacket method is most common in high power argon lasers.

Since the argon laser use arc discharge method to produce lasing where very high levels of excitation energy is needed. This high level of excitation produces extremely high currents through the bore or capillary tube. So the current density through the ore tube reaches several hundreds amperes per centimeter. It is that current which produces excessive heat. Now to resist this heat, the bore tube must be constructed from such material like beryllium oxide, graphite or tungsten. Usually the bore diameter is about 0.1125 centimeter.

The ionized gas in the bore tube contains many charged particles which force themselves away from each other towards the edges of the tube where they become cooled and thus lose the necessary energy for excitation. To solve this problem the magnetic coil or solenoid is placed around the tube and a magnetic field applied to the ionized gas. The magnetic field forces the ions back into the center of the tube where their energy can be retained.

The mirrors in argon laser are not directly connected to the laser tube because high temperature are associated with argon. Instead of direct connected mirrors, the Brewster windows are utilized at either end of the tube between Brewster window and HR mirror. There is inserted a device called prism.

In argon laser the power supply is also used which must be able to supply large amounts of current needed for lasing process and also provide enough voltage for the solenoids operation. The power supply is connected to anode and cathode of the tube.

Function of Argon Laser

We know that the active medium in an argon laser is a pure argon gas. The gas can reach to the excited state (necessary for lasing) through either direct excitation of argon atom or through further excitation of the already excited argon ion. In either process more of the argon atoms follow the two steps process that is atom-to-ion & ion-to-further-excited-ion. These two state processes has two results, 1st one is "intense or extremely high current discharge" and 2nd one is "the production of several wavelengths of energy".

Simplified energy level diagram for an argon laser can be shown as:

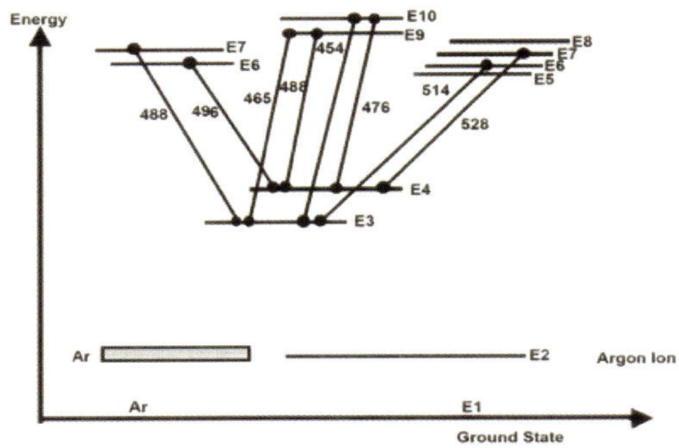

This diagram shows that a number of different wavelengths are produced by lasing argon gas. In all these wavelengths only two 488 nm (blue spectral region) and 514 nm (blue-green spectral region) are very important and predominant. All other wavelengths produced have considerably less power than either of these two.

As with all lasers, the argon laser also uses a pair of mirrors as feedback mechanism. However since argon laser is a higher gain system there for the percentage of reflectivity for the mirrors is less critical.

Applications of Argon Laser

- Argon lasers used in scientific research.
- Argon lasers used in medical application.
- An artistic displays and light shows also the argon lasers are used.

Advantages of Argon Laser

- Production of multiple wave lengths is the main advantage plus characteristic of argon as well as other ion lasers.
- Argon lasers produce high power output as compared to He-Ne laser.
- Argon laser is a higher gain system. Argon laser like He-Ne has very less divergence, typically about 1 milli radian.

Disadvantages of Argon laser

- The overall efficiency of argon laser is very less usually lies between 0.01% and 0.1%.

- Large amount of power requirement is also its disadvantage.
- Construction is very difficult.
- Cost of argon laser is not as low as He-Ne laser.
- Power supply of high voltages required, because due to solenoid there is extra burden on it.

Krypton Ion Laser

Krypton Ion laser is another important member of the Ion lasers category.

Argon lasers are though by far the most important members of this group, but krypton ion lasers are also in wide use. Ion lasers are gas lasers in which the stimulated emission process occurs between two energy states of an ion. The excitation mechanism of such a laser first must supply the necessary energy to remove an electron from the lasing atom to produce the ion, and then must supply additional energy to raise the ion to the appropriate excited state. These large input energy requirements result in low efficiencies for essentially all ion lasers. The energy levels of krypton ion lasers are similar to those of argon, and they also may be operated either on one line at a time or on several lines at once. A unique feature of krypton lasers is that, with the proper mirrors, they will lase on four lines that are red, yellow, green, and blue in color. This produces an output beam that is white in appearance and is uniquely suited to laser light shows. Krypton ion laser employs ionised krypton gas (Kr^+) and lases at a number of wavelengths (more than 10), most important being in the visible region of electromagnetic spectra.

Krypton ion laser and argon ion lasers are similar in construction and performance, with the argon system producing higher powers for longer lifetimes. Krypton-ion lasers are almost identical in construction and reliability to argon lasers. Krypton lasers emit at several wavelengths : in the visible range it emits at 406.7 nm, 413.1 nm, 415,4 nm, 468.0 nm, 476.2 nm, 482.5 nm, 520.8 nm, 530.9 nm, 568.2 nm, 647.1 nm, 676.4 nm. The argon laser has its strongest output at 514 nm (green) and 488 nm (blue). The krypton laser is known for its red (647 nm) and yellow (568 nm) output. The two gases can be combined to produce a laser with quasi-white all-lines output. Under some conditions krypton lasers can produce wavelengths over the full visible spectrum with lines in the red, yellow, green and blue. The 647.1 nm and 676.4 nm red lines, however, are the strongest and result in the best performance. The relative power of various lines is shown in this table.

Wavelength	Relative Power
406.7 nm	.036
413.1 nm	.07

415.4 nm	.02
468.0 nm	.02
476.2 nm	.016
482.5 nm	.016
520.8 nm	.028
530.9 nm	.06
568.2 nm	.044
647.1 nm	.014
676.4 nm	.048

Efficiencies of Krypton ion lasers is also low because of the large amount of input energy required to ionize atoms and excite them to the proper state. The most general expression for the output power of an ion laser indicates that the power increases approximately with the square of the current density. Implying that smaller bores produce higher powers for the same current values. Thus, ion lasers are constructed to have the smallest bore diameters possible without introducing excessive diffraction loss or erosion.

Krypton lasers, like argon ion lasers, may be operated at gas pressures from 250 to 500 millitorr, but the best operation is usually in the range of 300 to 350 millitorr. Specific systems are designed to operate best at a specific gas pressure, and the laser power supply is matched to tube characteristics at that pressure. If gas pressure is too low or too high, electrical characteristics of the tube may become incompatible with the power supply, resulting in failure of the tube to ignite or in unstable operation. The only two materials that are practical for such tubes are graphite and Beo. The discharge tubes typically have an inner diameter of 2 to 5 mm., and a length of 350 to 450 mm. Passing a high-current discharge through a tube, in general, excites ion lasers. The discharge is concentrated in a small-diameter bore at the center of the tube, which is where laser action occurs. An initial spike of a few thousand volts breaks down the gas, then voltage drops to 90-400 V while the current jumps to 10-70 A in the sustained discharge. An external magnet producing a magnetic field parallel to the bore axis can help confine the discharge to the bore. Krypton generates only 10%-30% as much power as argon used in the same tube. The strongest line of singly ionized krypton (Kr^{+}) is at 647.1 nm (red), but other lines in violet, yellow, green and red (416 nm, 530.9 nm, 568.2 nm, 676.4 nm, 752.5 nm, 799.3 nm) can produce up to one half or slightly less as much power, The violet lines are important for dye laser pumping; the red and yellow lines are important for displays. Doubly ionized krypton (Kr^{2+}) has three near-UV lines. Laboratory lasers have also been shown operating on Kr^{3+} lines, which have generated CW powers at 242 to 266 nm and wavelengths as short as 219 nm.

The output power of Krypton ion laser also depends on the magnetic field strength. If there is no field, current density in the center of the bore is reduced and more energy is being lost through collisions with the walls of the tube. An increase of magnetic field strength increases current density and output power. The optimum value of the magnetic field is in a range of 600 to 1,200 gausses.

When, a gas is subjected to a strong magnetic field, each of its energy levels splits into several closely spaced levels due to Zeeman effect and is proportional to the strength of the magnetic field. The magnetic field in ion lasers results in a broadening of the laser gain curve that is proportional to the applied magnetic field strength. At field strengths below about one kilogauss, this produces a broadened output spectrum with more cavity modes and higher output power. Above about one kilogauss, the gain curve becomes so broadened and flat that its edges fall below the lasing threshold and the output power and spectral lines will both begin to drop. Each laser line has an optimum magnetic field strength.

Krypton ion and argon ion lasers are very similar - they are both rare gas ion lasers, their basic principles of operation are similar, and the same basic hardware configuration and power supplies can usually be used. Differences are primarily in gas fill of the plasma tube and the mirrors/prisms for selecting the output wavelength. Sealed plasma tube with internal or external mirrors and high current regulated power supply. Combined Ar/Kr produces lines in red, green, and blue, and is therefore considered a 'white light laser'.

Normally, optics is selected to support the mission of the laser. For example, surgery wants only the blue lines; ophthalmology needs green, red, and yellow; Raman Spectroscopy needs 647 and 676 nm; laser shows use argon for blue, green, and violet, and krypton for red and yellow. Mixed gas lasers use optics selected for 55-60 % red, 20% green, and 20 - 25% blue and violet. To suppress a line, one of the optics is made more then 15% transmission for that line.

Since the krypton and argon lasing mediums have substantial gain at several spectral lines, a given laser can be set up to output on a single line or more than one at the same time depending on how the optics are designed and adjusted. The tube current also affects this to some extent as increasing the current will bring in progressively more lower gain lines.

A laser set up for multi-line operation will usually result in highest total output power but there are many applications like holography where a monochromatic beam is required. Single line operation is achieved by replacing the multi-line rear mirror with an intracavity prism assembly as mentioned in the case of Argon ion lasers. This assembly consists of an internal prism aligned to properly deflect the intra-cavity optical path to the High Reflector. Because of the dispersive properties of the prism, only one wavelength at a time will be properly aligned and produce lasing. The wavelength selector thus allows easy tunability and selection of any of the individual lasing wavelengths. As mentioned earlier using special optics sets coated to transmit more then 15% for the lines to be suppressed can also be used to implement single line operation, thus stopping them from oscillating.

Multi-line operation requires a set of mirrors with reflectivities designed to achieve laser operation for all the desired spectral lines. Any intracavity prisms are removed.

Beam quality is very high and one can achieve TEM_{00} mode Gaussian beam, for all lines using suitable optics.

White Light Lasers

The term 'white light laser' typically refers to one that is capable of producing a set of wavelengths which if mixed in the proper proportion can 'simulate' the effect of a white light source in full color displays and laser shows and also for some spectroscopy applications. However, they generally don't produce a broad spectrum like an incandescent light bulb.

Though under some conditions krypton lasers as such can produce wavelengths over the full visible spectrum with lines in the red, yellow, green and blue. However, the most common white light lasers are large frame ion types with a mixture argon and krypton for the gas fill.

White light lasers are now even available in air-cooled format. All use a mix of argon and krypton. Many are made for a roughly 60:20:20 ratio of red, green, and blue lines for proper white balance.

Applications

The applications of Krypton ion lasers are almost the same as those of Argon ion lasers. These include areas such as very high performance printing, copying, typesetting, photo-plotting, image generation, forensic medicine, general and ophthalmic surgery, laser shows for entertainment, holography, electro-optics research, optical 'pumping' source for other lasers, spectroscopy etc. Krypton lasers are also used in medicine for example for coagulation of retina.

Copper Vapor Laser

Copper vapour laser (CVL), invented in 1965 by scientist William Walter, is a three-level laser and uses copper vapours as the lasing medium. It produces green laser light at 510.6 nm and yellow laser light at 578.2 nm. It consists of a sealed zirconia tube filled with neon gas at a pressure of 25-40 torr. A solid block of pure copper metal is kept in the middle of the tube. Copper is heated above 1083°C (laser operating temperature is about 1650°C) to generate copper vapour, which is the active laser medium. High voltage is applied between the two electrodes at the end of the zirconia tube. As a result, the temperature rises inside the tube cavity to about 1400 - 1700°C, until the Copper evaporates, and the vapour pressure of the Copper is about 0.1 torr. During the laser operation only a small fraction of the copper atoms are ionized, and they move toward the ends of the tube because of electrical attraction. There, the vapour cools down, and transforms to solid metal, which reduces the quantity of copper in the tube. To continue the operation of the laser, it is necessary to replenish the copper metal in the tube after few hundred hours of operation. Electrons, accelerated by the high voltage applied to the electrodes, collide with the copper vapour molecules, exciting them into one of the available high laser energy levels. The lasers are self-heated such that most of the energy provided by the discharge current provides heat to bring the plasma tube to the necessary temperature.

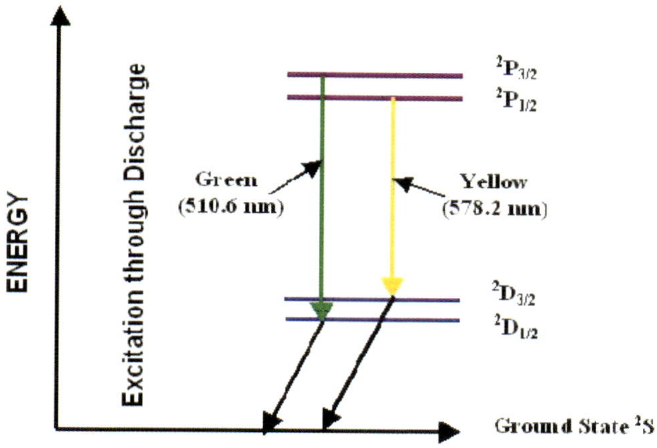

Energy Level Diagram of Copper Vapour Laser

Excitation occurs by electrons colliding with neutral copper atoms to excite them to the relevant laser-related energy levels. Inelastic collisions of electrons with copper vapour atoms causes excitation of the copper atoms, so the inversion population occur and laser oscillation due to electron transition from upper level ($P_{1/2}$, $P_{3/2}$) to the lower meta-stable level ($D_{5/2}$, $D_{3/2}$) take place at the 578.2- and 510.6nm, respectively. Two principal outputs having wavelengths of 510.6 nm (green line $^2P_{3/2}$ - $^2D_{5/2}$) and 578.2 (yellow line $^2P_{1/2}$ - $^2D_{3/2}$) are obtained. The lower laser levels (2D) are metastable leading to self-termination of the laser action. Upper and lower laser levels are shown in the energy level diagram.

The Copper Vapour Lasers (CVL) can be realized by two different ways. First one is the development of lasers using copper as such and the second one involves the vapours of copper-bearing compounds, mainly the copper halides, CuCl, CuBr, CuI. In case of copper based CVL lasers, Copper must be heated to 1400 to 1700°C in order to achieve a suitable vapour pressure. However, in case of CVL utilizing copper based compounds, it is possible to achieve a sufficient copper concentration for lasing in the 300 to 600°C range depending upon the type of compound. Typically Halide based lasers such as copper bromide (CuBr), copper chloride (CuCl) and copper iodide (CuI) lasers are necessary to be heated to 400, 500, 600°C, respectively. The use of copper halide instead of pure copper as the active medium not only reduces the requirement for the heating but also is much easier to handle in terms of materials involved for the resonator and cooling requirements for electrodes and windows or mirrors. The benefit of the copper halide however has the disadvantage, of a more complex power supply. To dissociate the copper from its halogen atoms, a double pulse power supply is needed that is capable of firing two HV pulses within the recombination time of the copper halide molecule. Typically two energizing pulses in quick succession are required, the first to dissociate vapour molecules, and the second to cause the dissociated ions to lase. The first pulse provides copper atoms by dissociating the halide. The second pulse is delayed until an adequate copper atom concentration has built up. The second pulse is a fast discharge pulse that pumps the copper atoms to the upper laser levels by electron collisions. For better performance, the laser requires fast excitation pulses of rise time of the order of 100ns or less. The time gap between dissociation pulse and excitation pulse should be dependent on how fast the chloride and copper atoms recombine to copper halide. So HV pulse power supplies operating in the kHz range require fast switching devices such as hydrogen thyratrons.

Features of Copper Vapour Lasers

- In order to have copper vapours, the metal needs to be at very high temperatures, so the tube is made of Alumina or Zirconia, which are high temperature resistant materials.
- The tube diameter is 10-100 mm, and it contains Neon gas at a pressure of 25-50 torr.
- Compared to the pure copper vapour laser (CVL) that requires one HV pulse for stimulated emission, the copper halide laser (CHL) needs two. Therefore the power supply design is complex. The delay time of the two HV pulses is in the range of 100 μs to 280 μs, depending on the halide used.
- The heating is done by three 600 W IR heating elements, which are arranged in 120° angle around the tube. Around the heating elements a ceramic housing forms the constant temperature oven. The oven contains also a temperature sensor and insulating material.
- An unstable resonator is normally used for a copper vapour laser. However, for short cavity, flat - flat can work as well. Usually Rear mirror is a high-reflector -mirror 99.8% and front mirror is an output coupler with reflector of 2-4%. Since copper vapour is high gain, in very large units a normal glass with 4% reflectivity can be used as an effective output coupler. For smaller lasers, a partially reflective mirror of about 2% may be used.
- Since the energy levels are separate, the two emission wavelengths (510 nm and 578 nm) operate simultaneously. A dichroic mirror as a spectral filter is used to isolate both lines of the laser from each other.

- The output power depends on length, diameter, excitation voltage, temperature and pressure in the cavity.
- The maximum average output power for CVLs ranging in tube bore from 10 to 60mm, is given by $P_{max, aver}$ (Watts) = LD/100 D is diameter in millimeters and discharge length L in centimeters.
- Though both wavelengths are produced simultaneously, the energy per pulse of green emission interestingly depends on the frequency of the applied electrical pulse, whereas it is independent of the same in the yellow emission.
- The green: yellow ratio is dependent on plasma tube temperature. In general, higher temperature results in less green and more yellow. Typical values are 1.5: 2.0 for green to yellow at discharge temperature around 1500°.
- Neon is the best buffer gas and next best is helium.
- Copper vapour lasers are generally flowing buffer gas systems, particularly for high power applications. However with the flowing gas design, the large diameter beam, their size and weight, copper rods have to be replenished quickly at regular intervals. Nevertheless small sealed tube lasers are also possible with power levels of 5 - 10 W with small diameter beams. The life of these portable lasers is of the order of 300 - 800 hours.
- Using singly ionized species of Cu, researches have also demonstrated copper vapour lasers that are CW (continuous-wave), i.e., not pulsed, and lase at deep ultraviolet wavelengths. These lasers can provide average UV powers of several mW and are potentially useful for analytical instruments and spectroscopy. CW copper-vapour laser efficiency is of the order of 0.1% relative to the discharge power.

Typical Parameters of Copper Vapour Lasers

- Average Power (W): Upto 200 W;
- Peak Power: 50 - 500 kW;
- Pulse repetition rate: 2 - 40 kHz;
- Green/Yellow ratio 1.5:1;
- Pulse width (ns) 5 - 50 nS;
- Efficiency (%) Greater than 1;
- Small signal gain go = 0.05 - 0.1 $cm^{-1;}$
- Saturation intensity = 9 - 12 $W/cm^{2;}$
- Beam Diameter: 5 - 15 mm;

- Divergence: 3 - 5 mrad;
- Tube diameter: 10 - 100 mm;
- Tube length: 50 - 150 cm;
- Warm up time: 45 - 90 min;
- Lifetime: 300 - 800 hours;
- Buffer gas: 25 to 50 Torr of neon;

Applications

- Pumping source for tunable dye lasers and solid-state laser materials such as Ti:sapphire to obtain pico second and femto second ultrashort pulses;
- High-speed flash photography and high-speed imaging with high spatial resolution and temporal resolution;
- Precision material Processing;
- Underwater applications;
- Holography;
- Particle imaging velocimetry;
- Spray Pattern Measurement;
- Flow Visualization;
- Photodynamic therapy and detection of forensic evidence;
- Laser beam can be absorbed by biological tissue components, which may be selectively destroyed. In oncology, the photodynamic therapy based on the effect of simultaneous photochemical reaction between an appropriate sensitize, laser light and oxygen, is used for a selective destructions of pathological tissues;
- Dermatology;
- Copper vapour lasers emitting light at 511 nm (green) ad 578 nm (yellow) have been useful for treating pigmented and vascular lesions, respectively;
- Nonlinear frequency conversion to the ultraviolet. Harmonic generation can produce 255 and 289 nm from the fundamental copper lines or tunable ultraviolet light from CVL-pumped dye lasers. So one can get pico-second and femtosecond pulses. They are particularly useful for studying Time-spatial resolved spectroscopy;
- High Resolution spectroscopy;

- Frequency doubled CVL can be used for fabrication of fiber Bragg gratings (FBGs);
- Copper vapour lasers has an important application in atomic vapour isotope separation (AVLIS) as a pumping source to excite tunable dye lasers.

Strontium Vapor Laser

Strontium vapour laser produces a high intensity pulsed wavelength having a wavelength of 430.5 nm. This wavelength falls in the blue-violet region of the visible region. The lasing medium is strontium metal gas. Laser action is observed when the strontium ions (Sr^+) transition between the two infrared states. The laser action of strontium was first observed by Deech and Sanders at the Clarendon Laboratory, in the year 1968.

In order to attain population inversion, helium or neon is used as buffer gas. A pulsed electric discharge is used to excite strontium. The lasing medium consists of a small amount of strontium gas that is maintained in a relatively high pressure helium buffer gas.

Laser Properties	
Laser type	Metal-vapour
Pump source	Pulsed electric discharge
Operating Wavelengths	430.5 nm

Applications

The applications of the helium selenium laser are limited. This laser is mainly used for scientific research. Some of the other applications of the strontium laser are in remote study of the meteorological parameters of the atmosphere.

Helium-cadmium Lasers

The helium-cadmium laser was discovered by William Silvast in 1966 at the University of Utah (Salt Lake City). The transitions in Helium-Cadmium laser are between energy levels of singly ionized Cadmium atoms, and there are about twelve lines. These wavelengths are in the shorter wavelength region, violet and Ultra-Violet (UV). The most prominent wavelengths are 441.6 nm and 325 nm. The helium buffer gas plays basically the same role for He-Cd as it does in the helium-neon laser. Excited helium in the discharge couples extremely well to the upper laser level of the cadmium ion. Helium-cadmium lasers typically have less than 10 Torr of helium constrained in a bore of the order of 1 mm in diameter, with vaporized cadmium at about 0.1% of the helium concentration.

The main mechanism identified for the laser transitions is a process known as penning ionization, in which highly excited helium atoms transfer their energy to cadmium in a way similar to the operation of He-Ne laser. However, in this case cadmium ions are produced in the process, instead of neutral atoms as in the case of neon, owing to much lower ionization potential of cadmium than of neon. That is why helium-cadmium lasers can be categorized either in metal vapor lasers since cadmium is a metal and the lasing action in helium cadmium laser occurs between energy levels of cadmium ions OR a gas laser since the properties of Helium-Cadmium laser are similar to those of helium neon laser, which is a neutral atom gas laser. Excitation energy for the helium-cadmium laser is provided by a direct-current discharge passing through the laser tube.

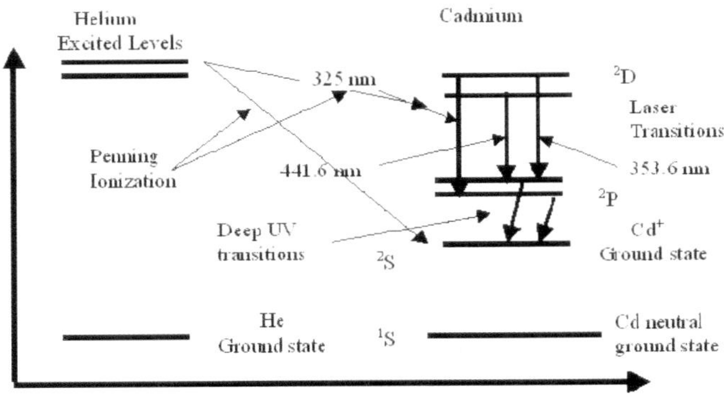

Simplified energy level diagram of Helium Cadmium Laser

Typical discharges are around 700 - 2000 volts, with current densities in the small-diameter bore (2 -mm) of the order of 3 - 5 amperes per square centimeter of cross section. Helium atoms in the laser gas absorb energy from the discharge and then transfer that energy to cadmium ions. The energy levels of cadmium and helium involved in the principal He-Cd lines are shown in the adjoining figure. The most prominent transitions, which can be easily, produced are 441.6-nm (blue transition) and the 325-nm (ultraviolet transition).

The most important energy transfer mechanism for the narrow-bore tubes mainly used for blue and ultraviolet lasers is Penning ionization. Penning ionization is a form of chemi - ionization, an ionization process involving reactions between neutral atoms and/or molecules. The process is named after the Dutch physicist Frans Michel Penning, who first reported it in 1927. Chemi-ionization is the formation of an ion through the reaction of a gas phase atom or molecule with an atom or molecule in an excited state and should not be confused with chemical ionization.

In Penning ionization, energy from an excited helium atom ionizes a cadmium atom:

$$He^* + Cd \rightarrow He + Cd^+ + e^-$$

The energy levels of cadmium and helium involved in the principal He-Cd lines are shown in the above figure. The 441.6-nm blue transition and the 325-nm ultraviolet transition, which are closely related, are the easiest to produce and the most widely available. Considerably more energy is needed to raise Cd+ ions to the upper laser levels of the red and green transitions which themselves are also closely related. In practice, different discharge conditions are necessary. The blue and ultraviolet lines can be generated in the positive column of the discharge, the region from the middle of a discharge to the anode where the electric field is not steeply graded and the electrons are not accelerated rapidly. The higher energy needed to produce the red and green lines is available only in the region near the cathode, where there is a larger change in electric field and a stronger electron acceleration. This difference means that special tube designs are needed to generate the red and green lines.

Because helium's lowest excited energy level is well above the ground state of Cd+, Penning ionization produces excited cadmium ions. The metastable helium excited states transfer energy to the levels of Cd+, which have lifetimes of the order of 100 nanoseconds (ns). The lower levels of Cd+ also receive some energy but are quickly depopulated because of their very low lifetime of the order of 1ns. The result is a population inversion between different states of cadmium, allowing laser action on either the 442- or 325-nm lines. The prism combination can be used to select the wavelength required for a particular application.

Cadmium, a metal, is solid at room temperature and for lasing it needs to be sufficiently heated to have required partial pressure of cadmium vapors in the discharge tube. Normally a uniform vapor is obtained within a chamber placing the vapor source either liquid or solid, and heating the entire chamber uniformly. In case of cadmium, it is not practical to heat the complete chamber to approximately 260°C because of the other components like mirrors, windows, electrodes etc. present in the chamber. Thus the challenge in making HeCd lasers operate continuous-wave (CW) is dealing with cataphoresis. This is the term given to the migration of the positively charged metal ions toward the cathode where they may condense, depleting the supply of vapor and contaminating tube components and optical surfaces. A critical breakthrough has been achieved to put cataphoresis to good use in distributing vaporized cadmium evenly throughout the discharge, which has made possible CW operation of the UV line. This is achieved by using cataphoresis to control the cadmium vapor distribution. In this process the cadmium metal is heated and vaporized at the anode (which is at positive potential) end of the discharge and is transported towards the cathode end of the discharge by the electric field acting upon the cadmium ions that are produced by the discharge current. The practical problem in Helium-Cadmium laser is to maintain homogeneous distribution of the metal vapor inside the electrical discharge tube. Once the vapors go out of the bore, it may deposit on cold surfaces. Thus it is important that its vapor remains at proper areas. In order to prevent coating of the windows with Cadmium, cold traps are put before the laser windows. The cadmium atoms then condense in a pocket near the cathode region. The laser species are mainly flowing down the capillary at a very slow rate: typically one to two gms per 5000 hours. The discharge current heats the gain region sufficiently so that cadmium does not condense in that region while it is being transported through the capillary.

The construction of helium cadmium (HeCd) lasers is far more complex than that of other helium based lasers. The laser tube contains a reservoir for cadmium and a heater to vaporize the metal. As a result, a heated filament cathode is often used in placed of the cylindrical tube that comprises a HeNe laser. Additionally, the laser itself needs to sustain a higher level of internal pressurization allowing the vaporized cadmium to remain in the tube. The lifetime of a specific helium cadmium laser is dictated by the amount of cadmium in the reservoir. Once the cadmium supply is exhausted, the tube must be replaced.

He-Cd laser tubes are more complex than those used for He-Ne lasers. In addition to often using a heated filament/cathode, they also include a reservoir for the cadmium metal and a heater to control its vapor pressure, a mechanism to add helium as needed to maintain correct pressure, possibly an overall heater and thermal insulation to control tube temperature, and various sensors inside the envelope to monitor these parameters for use by several feedback loops in the power supply. The power supplies are also correspondingly more complex with multiple feedback loops and power sequencing logic.

Some common features of he - Cd lasers are follows:

- The largest portion of the gas fill is helium (essentially 100 percent at startup), which sustains the electrical discharge and excites the cadmium vapor. Overall helium pressure is regulated via closed loop feedback. There may be a helium reserve with helium added as needed to maintain correct pressure. Unlike a He-Ne laser where there is no need to replenish helium over the life of the laser, a HeCd laser must periodically add helium to maintain the correct pressure. The partial pressure of helium is order of 1,000 times that of cadmium during operation.
- Like He-Ne lasers, He-Cd laser tubes also have a thick-walled tube with a small hole in it where the main discharge and laser gain actually takes place. But in He-Cd lasers, this may be made of fused silica or quartz instead of glass due to the higher current and bore temperature. The discharge current through this using the process cataphoresis assures that Cd vapor only travels down the bore to the cadmium condenser near the cathode, and not to the anode area and mirror where it could condense.
- In addition to the gas fill of helium, there is a cadmium reservoir and wrap-around resistance heater to maintain a specific (closed loop controlled) Cd vapor pressure in the tube. At startup, the cadmium is at ambient temperature

with negligible vapor pressure. A minute or two after the heater is energized to increase the temperature to around 260°C; cadmium pressure begins to increase noticeably. Feedback may be based on tube voltage or simply Cd temperature. During warm-up, residual cadmium remaining in the bore from previous operation may vaporize and momentarily produce some lasing, which then disappears until the cadmium reservoir begins to approach operating temperature. After the laser operation, cadmium is directed to end up in a generally cool place and may have a magnet to help direct the Cd.

- Overall wall-plug conversion efficiency for blue emission from a He-Cd laser runs from 0.02 to about 0.003 percent, higher than argon ion but below helium-neon. Efficiency of He-Cd on the ultraviolet line is about one-half that in the blue.

Applications

- Lithography.
- Stereo lithography in which the ultraviolet laser is used to make computer-generated models in a plastic material.
- The blue wavelength is used for printing on photosensitive materials.
- Flow cytometry.
- Making CD masters.
- Microchip inspection.
- Fluoroscence analysis.
- Diffraction grating fabrication.
- Spectroscopy.
- Nondestructive testing.
- Laser tumor cancer diagnoses.

Important Properties of He-Cd Laser are Follows

- Laser wavelengths: 441.6, 353.6, 325 nm;
- Small signal gain coeff : 0.2 - 0.3 m^{-1};
- Saturation intensity: 0.4 W cm^{-2};
- Gas mixture : He:Cd :: 100:1;
- Gas pressure : 5 - 10 torr;

- Laser gain medium length: 20 - 200 cm;
- Output power : Upto 200 mW;
- Mode : TEMoo or Multimode;
- Life time : Upto 6000 Hrs;
- Starting voltage: 10 kV DC;
- Operating voltage: 700 - 2000 V DC;
- Operating current : Upto 100 mA;
- Beam diameter :0. 3 mm for single mode and 2 - 3 mm for multimode;
- Divergence 1 - 2 mrad;
- M^2 : 1.3- 1.5 for single mode and 4 - 5 for multimode;
- Coherency length: approx. 30cm;
- Overall wall-plug conversion efficiency : 0.003 - 0.02 percent.

Nitrogen Laser

Nitrogen laser was developed as one of the first ultraviolet lasers in 1963. These lasers are convenient and economical sources of short, nanosecond, ultraviolet (337.1 nm) pulses. The gain medium is nitrogen molecules in the gas phase. The nitrogen laser is a 3-level laser: the upper laser level is directly pumped, imposing no speed limits on the pump. These lasers are based on a fast electrical discharge through N_2 gas. Traditional designs require vacuum pumps and flowing gas. Smaller sealed tubes, a more recent variant, are much more convenient to handle. These lasers are based on pure nitrogen, nitrogen-helium mixture, and sometimes even simply air. Emission typically occurs at 337.1 nm. The high gain leads this to relatively efficient super-radiant pulsed molecular laser. Super luminescent emission implies that laser can works even without a laser resonator. As a super luminescent source, it has a very low temporal coherence. It contains a laser gain medium, which is excited in order to emit and then amplify luminescent light. Since it has a large emission bandwidth of about 0.1nm compared to most other lasers: that is why its temporal coherence is very low. Nitrogen lasers are relatively easy to build and operate, and have been made by many hobbyists without refined laboratory equipment.

Although nitrogen lasers have been operated over a range of partial pressures from a few Torr to more than 1 atmosphere, it is common to divide them into two categories: low pressure and atmospheric pressure. The upper-state lifetime of nitrogen is inversely related to pressure; it is approximately 40 nsec at pressures of a few Torr, decreasing to around 2 nsec at 1 atmosphere. However, the lower laser level has much longer lifetime thus causing the laser as self-terminating. This effect seriously limits both the duration of the pulse and the efficiency of the laser. There are two versions a nitrogen laser, which may be constructed. The first approach is a low-pressure design - it is more 'traditional' and requires a vacuum pump. It produces a pulse of 5 to 10nS in duration, which has a cross-section of about 10mm by 2mm. The power output and beam shape are quite suitable for pumping a dye laser. The second approach is a TEA version of the nitrogen laser. TEA lasers (for Transverse Electrical-discharge at Atmospheric pressure) do not require a vacuum system at all as they operate at atmospheric or even greater pressures. Either laser, if built efficiently, can even operate using regular air as the lasing medium. In the case of the TEA laser using air, no gas housing is required. Although the idea of a TEA laser is tempting, since no vacuum pump required, construction of this type of laser is bit difficult than the type employing a vacuum pump. Ultra-fast discharges are required for the TEA laser (10 times faster than a 'normal' low-pressure nitrogen laser).

Nitrogen lasers are capable of generating peak power output of few megawatts, with a pulse width of 10 nsec, at rep rate 1000 Hz. at 337.1 nm in the UV region, with average energy of hundreds of milliwatts. Since the life times of it's upper and lower laser levels are about 40 n sec and 10 microseconds respectively, CW operation is not possible. The energy level diagram of the Nitrogen laser is shown below.

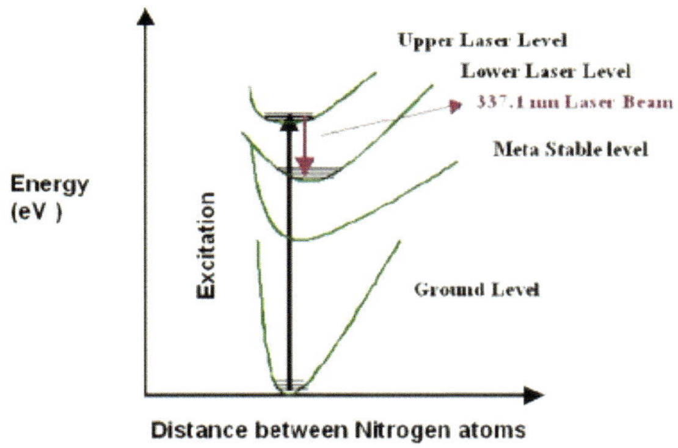

Simplified energy level diagram of Nitrogen Laser

A fast high-voltage discharge populates the upper laser level, an excited electronic state with 40-ns lifetime, which emits at 337.1 nm when it drops to the lower laser level as shown in the figure. The transition is a vibronic one, in which both electronic and vibrational energy levels change, making it broadband. The lower level has a 10- microsecond decay time, much longer than the upper level, and drops to a metastable state with a lifetime of the order of seconds. This implies that laser action is possible only if a fast electric discharge populates the upper N_2 level very efficiently. Pumping is normally provided by direct electron impact; the electrons must have sufficient energy, or they will fail to excite the upper laser level. Typically reported optimum values are in the range of 80 to 100 eV per Torr-cm pressure of nitrogen gas.

A fast strong electrical pulse does this where electron collisions cause the preferential population of the upper energy band first. After about 20nSec, the population of molecules at the upper laser level starts decaying to the lower laser level where it will stay because of much longer life time thereby quickly ceasing the laser action after the electrical pulse. Nitrogen lasers are hence self terminating. The pulse length of the low-pressure Nitrogen laser, then, is limited by the lifetime of the upper laser level i.e. 20nS. After this time, population inversion is no longer possible since half of the molecules in the upper energy state have decayed to the lower state.

The lifetime of the upper-lasing level is dependent on the pressure in the laser tube. As pressure rises, the lifetime shortens according to:

$$t = 36/(1+12.8*p(bar))\ ns$$

or

$$t = 36/(1+p(torr)/58)\ ns$$

There is 40 ns upper limit of laser lifetime at low pressures and the lifetime becomes shorter as the pressure increases. For TEA nitrogen laser where the pressure is 760 torr (one atmosphere) the lifetime is about 1 - 3 ns.

These features result in making N_2 lasers very high gains during its short pulses; small signal gain is typically 1-2 cm^{-1}. This avoids the need for a high-quality resonator. The laser gain is extraordinarily high, so much so that a single pass of light down the laser tube amplifies radiation enough to produce a powerful output beam. No mirrors are required for this laser - this is why, it is called a super radiant laser. The output beam simply passes through a thin glass microscope slide to exit the laser tube. This also eliminates mirror alignment problems mostly encountered with most other types of lasers. Though all nitrogen lasers can operate in super-radiant mode without cavity mirrors, but output can be enhanced significantly typically more than double, with a simple cavity with 100 percent reflective rear mirror. Further the use of rear mirror ensures that the output is emitted from the opposite end. Use of a high-reflectance rear mirror not only increases the peak power output but also decreases beam divergence, and improves beam homogeneity. Cavities are normally are 15 to 50 cm long.

The greatest challenge is the need to deliver a pulse of 15 to 40 kilovolts (kV) with rise time less than 10 ns. Previous designs involved the use of spark gaps, however, in recent designs, high efficiency thyratron switch provides the high energy required for pumping.

some other important features of nitrogen lasers are:

- Strong laser emission is there at 337.1 nm, whereas weak emission has also been reported at 357.6.

- Wall-plug efficiency of N_2 lasers is typically of the order of 0.1 percent.
- Pulse length dependent on both gas pressure and the discharge circuit. Pulse lengths range from about 300 picoseconds (ps) at atmospheric pressure to about 10 ns at 20 torr. Repetition rates range from 10 to 100 hertz (Hz) in most of the lasers, though 1000 Hz has also been demonstrated.
- A number of different geometries for N_2 laser resonators can be employed. Usually a channel of narrow, rectangular cross section is used as a resonator. Though both transverse and longitudinal excitation has been used: but the most common technique is to excite the channel by electrical discharges transverse to the direction of coherent light output, i.e., the optical axis. Glass plates form the top and bottom of the channel. Long electrodes, sometimes called distributed electrodes, are sandwiched between the plates to form the sides of the resonator. The beam, like the channel itself, is of rectangular cross section.
- The energy storage/discharge circuit of an N_2 laser generally consists of five basic elements: External high-voltage power supply, Capacitor bank, Fast switch like thyratron, Impedance-matching network and Laser channel.
- Energy is stored in a capacitor bank charged by the external power supply. A trigger pulse then is applied to the grid of a hydrogen thyratron used for switching. When the thyratron conducts, the high-voltage pulse travels through the transmission line and appears across the electrodes of the laser channel. This results in a rapid discharge that excites most of the nitrogen molecules in the active medium. The excited nitrogen has an extremely high gain and lases for a few nanoseconds. This produces a high-peak-power pulse, but quickly destroys the population inversion.
- Beam profile is not so good. Low-power nitrogen lasers have beams as small as a millimeter or two across. Higher-power lasers typically have oval or rectangular beams up to 6 by 32 mm across. Beams derive this shape of oval or rectangular in cross section because of discharge cavity shape. Divergence ranges from 1 by 2 mrad to 4 by 6 mrad.
- Standard-purity nitrogen is adequate because N_2 lasers are not very sensitive to impurities. The nitrogen supply can be either of a standard gas grade, supplied through a gas cylinder, or from liquid nitrogen. Even some studies have shown the use of just air, which is 78% nitrogen. However, the efficiency is low in the case of air.
- The electronics is a circuit composed of a spark gap and a capacitor. The speed of this circuit is increased by reducing the inductance of the circuit.
- The ratio of electric field to pressure, E/p, of the gas is an important factor. Nitrogen lasers have been reported to operate over a wide range of E/p values

however it is generally acknowledged that the optimal E/p is around 120 to 300V/cm-torr. For a low-pressure design typically 100 torr, with a 1cm gap the operating voltage is of the order of 15kV.

Properties

- Wavelength 337.1 nm.
- Spectral bandwidth 0.1 nm.
- Pulse width (FWHM) < 3.5 ns.
- Pulse energy upto 300 μJ.
- Rep rate upto 100 Hz.
- Beam size 3 × 7 mm.
- Small signal gain: 1-2 cm^{-1}.
- Saturation intensity; 50 KW/cm^2.
- Beam divergence 4 × 6 mrad.
- Long tube life - 10^8 shots per fill.
- Spark gap or DC heated thyratron triggered discharge.

Applications

- Pumping source for dye lasers.
- Measurement of air pollution using LIDAR.
- Time-of-Flight Mass Spectrometry.
- DNA Sequencing.
- Laser Ablation.
- Production of fast, dense pulse of photoelectrons for materials testing.
- Biomedical diagnostics.
- Study of fluorescence effects.
- Studies related to Raman scattering.
- Measurement of particle size by light scattering.
- Use in optical coherence tomography (OCT) in the medical sector.
- For device characterization such as gyroscopes and fiber optic sensors.

Excimer Laser

An excimer laser is an ultraviolet (UV) laser that uses a compound of noble gases, halogen, etc. as its laser medium, typical examples being ArF excimer lasers (wavelength of 193 nm), KrF excimer lasers (wavelength of 248 nm), XeCl excimer lasers (wavelength of 308 nm), and XeF excimer lasers (wavelength of 351 nm).

Excimer lasers have the characteristics of being able to oscillate at an exceptionally high efficiency for lasers in the ultraviolet range and enabling comparatively compact hardware, and are applied in a variety of fields such as industry and medicine (vision correction surgery such as LASIK).

Examples of Excimer Laser Application

The high photon energy and extremely high peak power of the ultraviolet light emitted by excimer lasers are very attractive, and applied research is being carried out in a variety of fields to see whether this strong light can be utilized. For example, there are marking of materials that cannot be marked by CO_2 lasers or YAG lasers, aberration processing of macromolecular films, semiconductor thin films, etc., surface reforming of resins such as Teflon etc., semiconductor manufacturing processes such as doping, deposition, and hole processing for TSV.

One of the most successful examples among these is as a light source for lithography in semiconductor fabrication.

Also, together with the increase in definition of liquid crystals in recent years, laser annealing for liquid crystal manufacture has been attracting attention. This is an indispensable process in the manufacture of polycrystalline low temperature poly-silicon (LTPS).

Excimer Laser Key Technologies

High Voltage Pulsed Power Circuits–MSC and NovaPowerSwitch

The “heart technology” of any excimer laser is its electrical discharge circuit including high voltage switch, capacitors for energy storage and transfer, pre-ionization arrangement and electrodes. The first excimer lasers did use thyratrons as fast high voltage switches in rather conventional electrical circuits. The pulsed power circuit must switch tens of kilo- volts on a nanosecond time scale. The thyratrons used as the switching element in these power supplies represented a lim- iting factor for system lifetime especially if higher duty cycle of the laser operation was needed. A breakthrough in excimer technology was accomplished in 1983 when Lambda Physik’s scientists invented the magnetic switch control (MSC) for transferring the electrical energy efficiently into the laser chamber. MSC did prevent current reversal in the pulsed power circuit and allowed optimum adaptation of the thyra- tron switch to other high voltage components. This resulted in an increase in lifetime of the most expensive components of the laser–thyratron and laser tube–by more than one order of magnitude. Thousands of lasers have been installed with MSC technology with lifetimes of the thyratrons seen up to 4 billion pulses at some of them.

The development of very high repetition rate excimer lasers in the mid and late 90's however met the limit of the MSC technology. At 2 kHz repetition rate, for example a thyratron would survive only 23 days or less. The key to solve the lifetime issue was solved by completely eliminat- ing the thyratron and replace it by semiconductor switches as for example thyristors. In 1997 the development of the NovaPowerSwitch–a pulsed power circuit with low voltage semiconductor switch combined with magnetic isolator (MI) and pulse transformer to achieve the required discharge volt- ages could be introduced into the first industrial excimer lasers. The lifetime of a NovaPowerSwitch circuit is basically unlimited–determined only by the aging of the semiconductor components.

NovaTube and DuraTube Technology

Internal corrosion has long been one of the most serious problems limiting excimer laser performance and reliability. Over time, the corrosive gases used in excimer lasers chemically react with the laser tube and its components. This process has several negative effects on laser operation. The resulting contaminants quench laser action, thereby reducing laser operating efficiency, which can significantly limit the lifetime of a gas fill. Reduced efficiency also negatively impacts laser beam quality and pulse energy stability. Equally important, corrosion limits overall tube lifetime and increases the frequency of routine optics component cleaning and replacement.

Another milestone in the excimer technology could be achieved by the development of the NovaTube technology in 1994, an all metal/ceramic laser chamber construction that significantly reduces corrosion. Specifically, all insulators and high voltage feed throughs in the laser chamber are made from corrosion resistant high-density ceramics. The metal parts of the laser tube are of special carbon- and silicon operation, even at low concentrations at the ppm level. After clean room assembling of the NovaTube a multistep passivation process eliminates all traces of contaminants including water from the tube components, ensuring that exposed surfaces will remain inert. NovaTube technology did enhance the gas lifetime of the excimer laser by more than 10 times, even at the very short 193 nm wavelength. In 1997 NovaTube was improved by implementing a soft pre-ionization technique of type of a barrier-discharge corona giving further enhancement of both the gas and, even more important, laser chamber lifetime. However, also with NovaTube technology the laser chamber lifetime is still dependent on the rated output power of the laser. As a rule of thumb is valid the lower the power the longer the chamber lifetime. In 1998 for the first time a chamber lifetime above 10 billion pulses could be demonstrated for a 10W KrF NovaLine lithography laser operating at 1 kHz.) And also the high power laser LAMBDA 4308 with 200W output at 308 nm17) exhibits tube lifetimes in excess of 2 billion pulses–this is more than 6 months of operation in a typical production environment.

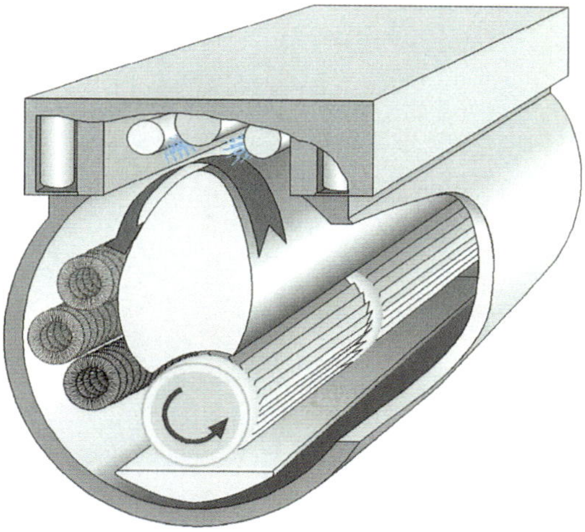

Fig: NovaTube excimer laser chamber

Extending NovaTube, in 1999 another laser tube technology was developed which matched the experience from NovaTube with the requirements of most efficient operation and longest lifetime at the 157 nm wavelength of the F_2 laser– the DuraTube technology. DuraTube, which today is used in all Lambda Physik 157 nm laser employs materials and preionization arrangement specifically developed for F_2 lasers. With DuraTube technology another world record in gas lifetime at 157 nm could be achieved in 2000–a NovaLine F1020 lithography laser did operate more than 1 billion pulses on a single gas fill.

Halo Safe Gas Generators

The storage and handling requirements of the halogen gases used in excimer laser operation make an additional investment necessary to install excimer lasers–the gas installation. While installation requirements for the rare gases (Ar, Kr, Xe) and the buffer and flushing gases (Ne, He) are simple, the gas installation for the halogen gas supply according to the usual safety regulations is more demanding. At installations of several lasers, the cost of the halogen gas installation divided by the number of lasers operated, seems reasonable. At single excimer laser installations however, the cost of the gas and safety installations may contribute to a significant portion of the total cost of the laser system. The final solution of this problem would be a sealed-off excimer lasers which does nod need an external gas supply. A first step into this direction was introduced in 1994–the HaloSafe halogen gas generator system which is a sealed module, either built-in the excimer laser or externally connected to the laser. The two versions of the gas generator produce either HCl or F_2 on demand, from solid, inert materials just before a new gas fill of the excimer laser is necessary.

As example we explain here the principle of the fluorine generator: By heating of the complex salt $K_2NiF_6.KF$ pure fluorine is generated according to the reaction:

$$2K_2NiF_{6.}KF \rightleftarrows K_3NiF_6 + F_2$$

The fluorine generation occurs at temperatures above 300 C, below 300 C fluorine is adsorbed again in the salt. To generate the necessary amount of fluorine for a single laser fill the Halo Safe generator is heated computer controlled. As soon as enough fluorine is generated the amount is injected into the laser chamber and the Halo Safe reactor is cooled down. The excessively generated fluorine in the reactor is adsorbed by the complex salt and no gaseous fluorine remains in the reactor making the reactor a safe device.

However, gas produced with the fluorine or HCl generator is more expensive than that from the halogen bottle of high purity gas suppliers. Calculating the total amount of money needed to operate the excimer laser and taking into account local safety regulations, this technology is specifically favorable for medical and scientific installations.

Bandwidth-Narrowing Technologies

In the second half of the 80's a large push towards the edge of technological possibilities of excimer lasers came from their intended use in semiconductor chip manufacturing and the moving target specifications for the laser parameters. While early estimates did forecast the need of 248 nm excimer lasers for microlithography at critical chip dimensions of 1.0 μm, the actual start of production was at 0.25 μm dimensions in 1998. This shift in time was mainly caused by the immense improvement in resolution of Hg i-line microlithography steppers with higher numerical aperture lenses and sophisticated mask technology. For the excimer lasers as light sources the progress in the established lithography technology resulted in the demand of narrower spectral laser bandwidth, higher spectral purity, significantly improved energy stability and higher repetition rate. The required spectral characteristic of the excimer laser output is determined by the imaging lens of the deep UV scanner system which, at higher NA for higher resolution, tolerates only an ultra-narrow laser spectrum.

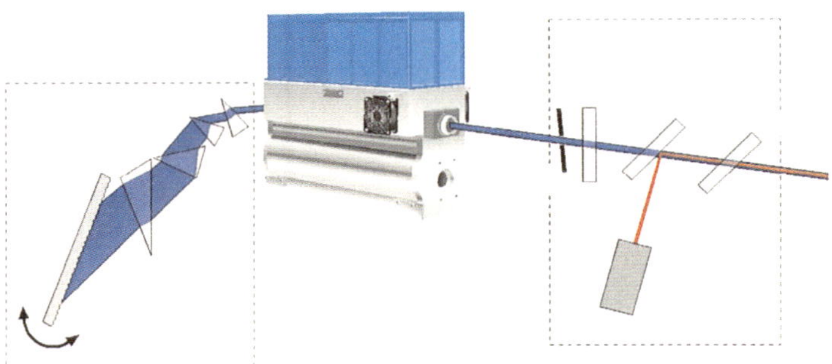

Fig: Narrow bandwidth resonator with optical elements as prisms and grating (left), laser tube with pulsed power module (center) and outcoupling mirror and beam diagnostics (right)

To generate narrow-bandwidth radiation from excimer lasers resonator the principles of the dye lasers invented independently by Fritz Schäfer and Peter Sorokin were applied by Lambda Physik researchers.) similar as in the early EMG 150 narrow-bandwidth oscillator. Early attempts of bandwidth-narrowed lithography excimer lasers did actually consider oscillator-amplifier configurations as that of the EMG 150. But it turned out that the cost of the laser system would be too high. The reliability needed for the use in semiconductor fabs seemed unlikely to be achievable. One of the main challenges for narrow-bandwidth excimer lasers for lithography was therefore the development of UV optical components including coatings with the needed surface quality and efficiency to achieve the required power levels from a single excimer oscillator. In the mid 80's Lambda Physik scientists started to work closely together with specialists in manufacturing of optical components for the UV range to accomplish this goal. The result was the first lithographytype excimer laser emitting at 248 nm–the Lambda 248 L. It had adopted exactly the dye laser resonator design principles of Schäfer and others which later have been applied also to the ArF excimers. The cavities of lithography lasers include highly efficient line-narrowing elements such as high-resolution optical gratings and etalons allowing to select bandwidths of the radiation of below 1 pm in the early.

With the use of lithography lasers in chip manufacturing the lifetime of the optical components under the high UV power densities occurring in the excimer laser resonators became a major issue. In addition to continuously improving the UV radiation resistance of the optics, resonator configurations were needed which reduce the power load on the optical elements. As result of a joint R&D effort between Lambda Physik and Matsushita Research Institute of Technology in Japan, the lifetime of the optical components in high power UV narrow-bandwidth cavities could be significantly increased employing the polarization coupled resonator principle PCR.) PCR significantly reduces the power load on the critical resonator elements and increases their life-time.

Recent Trends and Progress in Excimer Laser Technology

General Requirements for Excimer Lasers

Since 1977 the requirements from the various industrial, medical and scientific applications as well as technical possibilities in excimer development determine the trends of excimer laser technology. Different applications have different requirements. For example, for a R&D application in first line the parameters of the excimer laser must fit: energy, bandwidth, wavelength, energy stability or average power or repetition rate or pulse duration. A new energy level, higher repetition rate or another bandwidth from the excimer laser opens new research fields–with results researchers have in mind, but nobody may predict exactly. For industrial use the required parameters of the excimer laser usually have been fixed in a R&D phase (at least on a years basis). For use in production it is assumed the required laser parameters are met. Therefore, of higher importance is throughput, cost of ownership (CoO) and maintenance intervals since these determine the profitability of the specific application. Similar with medical application–installation cost, operating cost, safety requirement cost and maintenance cost have to be calculated carefully against the number of treatments expected.

Taking the requirements of all excimer laser applications together we see the main trends over the years as following:

- Continuous reduction in cost of ownership (CoO)–which includes easiest (or better no) maintenance, highest component lifetime and long exchange intervals, long gas lifetime with a long-term trend to sealed-off operation of the excimer laser (no gas installations are needed anymore), fully automated and software supported operation of the excimer laser by the main computer controlling the application result as well as remote control and diagnostic capability via intranet, internet or telecommunication.
- Shorter wavelength–as needed for high resolution lithography down to the 70 nm and may be to the 50 nm node as well as micromachining of difficult materials with highest resolution.
- Higher repetition rate operation–as needed for lithography to achieve the required illumination dose stability and lithography lens lifetime, for fiber Bragg grating manufacturing, for mask inspection systems for chip manufacturing, for micromachining with serial instead of parallel approaches, for medical use in refractive surgery with the customized ablation.
- Higher stability of the laser parameters as energy, beam profile, divergence, pulse duration, and peak intensity–for all applications.
- Higher laser power–to achieve the required throughput and/or the required size of the area of material which can be machined simultaneously.

Cost of Ownership Considerations and Shorter Wavelength Requirements

The continuous reduction of the installation, operation as well as maintenance cost is one of the main requirements for the use of excimer lasers in industry or medical environment. To accomplish easy maintenance and low mean time to repair (MTTR) numbers, Lambda Physik had developed a new series of industrial lasers in 1996–the NovaLine series. While engineered in a fully modular way ensuring shortest exchange intervals of consumables or failure components, the NovaLine series takes advantage of using the same modules as electrical supply components in the control cabinet, high voltage power supplies, on-board computer, gas handling system, cooling circuit, energy detection and stabilization hardware as well as total system software for all the different NovaLine models for various applications.

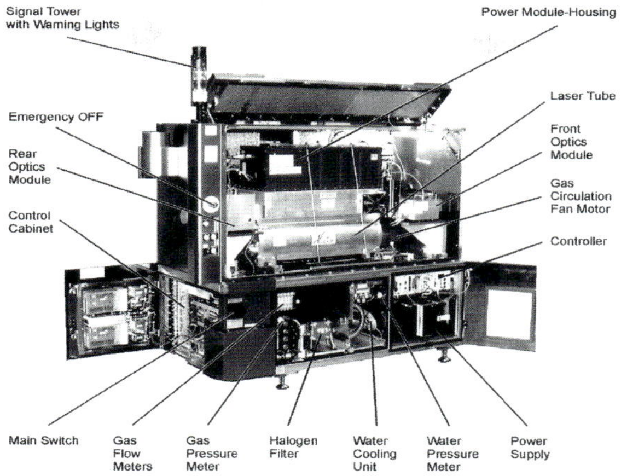

Fig: Modular designed industrial lithography and micromachining laser NovaLine with open covers showing the main laser modules

NovaLine is currently one of Lambda Physik's main industrial laser lines for microlithography and micro fabrication. One of the first Lambda Physik lasers for the use on the production floor of chip manufacturers, for example, was the NovaLine 248-0.8. The extremely narrow spectral width of the laser output of less than 0.8 pm FWHM at average power levels of 10W at 1 kHz was achieved using the newly invented polarization coupled resonator. For ink-jet nozzle drilling the NovaLine 100 is applied–a 100W KrF laser.

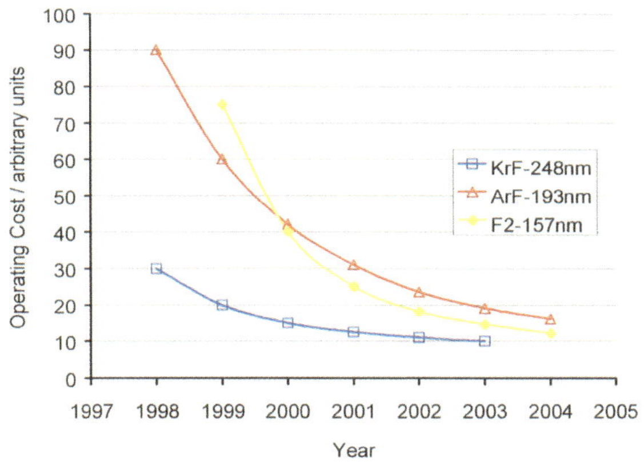

Fig: Development of CoO of 248 nm, 193 nm, and 157 nm lithography lasers

Cost of ownership calculations and projections are one of the driving forces in new developments of industrial excimer lasers at Lambda Physik. As one example representative for the whole industry, the historical and projected development of operating cost for 157 nm lasers for lithography. Specifically the projections of cost are influenced to a large extent by the changing specification requirements of the semiconductor industry. For example, the ArF excimer laser scanner technology which was earlier thought to be needed at 0.18 μm dimensions is currently in competition with the KrF technology for 0.13 μm chip production. Cost of ownership comparable to the 248 nm lithography wafer scanners is required. First ArF excimer laser scanners using the NovaLine A1020 and A1010 had been installed at semiconductor manufacturers for process development of 193 nm lithography earlier in 1999.

It is obvious that 157 nm lithography lasers must be competitive in CoO with 193 nm technology as otherwise the 157 nm lithography would not be profitable. Based on 20 years of experience with 157 nm lasers in a variety of applications, Lambda Physik had developed the worldwide first lithography F_2 laser, the NovaLine F630, in 1998. These lasers together with their successor NovaLine F1020 with DuraTube technology, a 10W single-line laser operating at 1 kHz, have been used immediately for material characterization, optics and resist development as well as lithography scanner development. Laser discharge chamber, solid-state pulsed power module and laser resonator optics have been optimized for laser emission with spectral bandwidth of about 0.6 pm. Major programs in lifetime enhancement of the expensive modules of 157 nm lasers did run and are running at Lambda Physik to follow or undergo the projected CoO.) Long term and durability tests of the 157 nm lithography laser system have been carried out comprising multi-ten billion exposure tests of individual components. Data from these tests as well as data obtained at several installations reveal lifetimes of the laser chamber in excess of 3 billion pulses and optics lifetimes above 2 billion pulses. Further programs are ongoing to reduce CoO until the introduction of 157 nm lithography into chip production.

Higher Repetition Rate Requirements

The requirement of higher repetition rate comes from several areas in industry and medicine. For lithography, laser repetition rate is at 4 kHz for 193 nm production tools. Medical application needs currently 200 Hz with 500 Hz expected and 1 kHz may be needed. Fiber Bragg grating manufacturing requires for certain processing repetition frequencies as high as possible. One of the demands fueling the development of high repetition rate excimer lasers is the need of higher dose energy stability in illumination and micro fabrication applications. For statistical reasons, it is better to have the total energy needed distributed among many pulses with low energy instead of having the entire energy in a single pulse, for example. High repetition rate is then necessary to achieve the required throughput.

Lambda Physik has lasers running at above 4 kHz to meet the requirements of the future. Years ago, when starting the discussion about 193 nm lithography in 1995, it was unbelievable this excimer laser output performance could be achieved.

Higher Power and Stability Trends

Applications as polycrystalline-silicon thin film transistor (TFT) annealing, ink-jet nozzle drilling, printed circuit board drilling, and lithography have been the main driving applications for higher average power and better pulse-to-pulse and long term energy stability for industrial excimer lasers.

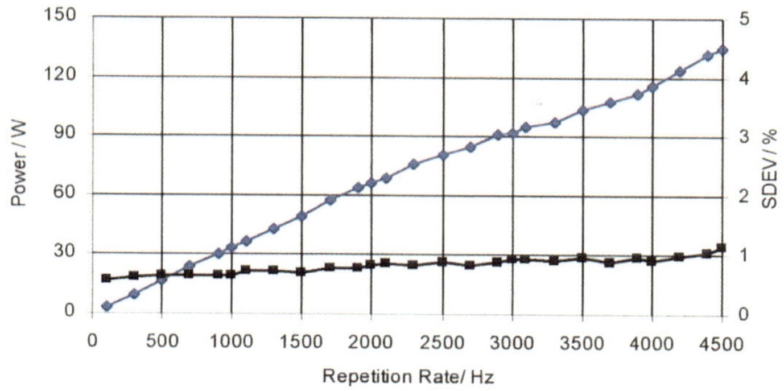

Fig: High repetition rate performance of an 193 nm lithography type laser with 120 W output power at 4 kHz

For example, in strategic alliance for the thin film transistor (TFT) LCD manufacturing Japan Steel Works (annealer), Micro Las (optics manufacturer) and Lambda Physik (laser manufacturer) have developed the first 308 nm excimer laser annealing system several years ago. The 308 nm wavelength has been shown to be well matched to the annealing process and to give significant advantage with respect to the optics lifetime in the annealing system. While the early systems have been used for process development and small scale manufacturing of active matrix LCDs, today TFT annealing systems are on the production floor since 1998. The necessary yield for high volume production sets stringent demands onto the energy stability of the laser. Increasing sizes of the displays require higher pulse energy and higher average output power.

In 2000, as successor of the TFT annealing laser LAMBDA 4000 the next generation of industrial lasers for excimer laser annealing, Lambda Physik's LAMBDA STEEL was introduced to the market with remarkably improved energy stability of better than 1% standard deviation at the high average power. The LAMBDA STEEL 1000 allows operation at 1 J pulse energy–at a 300W level the only commercial excimer laser available today.

The making of larger active matrix LCD displays will require more power from the excimer laser in the future–with demonstrated 1 kW excimer power in the research laboratory Lambda Physik is taking this challenge.

Argon Fluoride Laser

The argon fluoride (ArF) laser is a type of excimer laser. Excimer lasers (short for "excited dimer") consist of a mixture of noble gas such as xenon, krypton or argon and a halogen gas such as chlorine or fluorine, which when subjected to high pressure and electrical stimulation emit stimulated ultraviolet (UV) light in the range of 193 nm.

During high-voltage electric discharge, an excimer gain medium is pumped with current pulses to form excimers. The excimers in turn dissociate rapidly after spontaneous emission to prevent reabsorption of the generated radiation.

The UV light emitted by the ArF laser is beyond the visual range of human eye, and hence it is essential to use UV goggles to protect the eyes and skin as well from the carcinogenic properties of the UV beam.

Laser Properties	
Laser type	Gas
Pump source	Electrical discharge
Operating wavelength	193 nm

Applications

Organic compounds and biological matters easily absorb the UV light emitted by the ArF laser. This laser splits up the molecular bonds of tissues, which later disintegrates into the air via ablation in a controlled manner, instead of burning.

Therefore, the ArF lasers have a potential to eliminate fine layers of surface material without heating. This property enables ArF excimer lasers to be commonly used in high-resolution photolithography machines, a technology that is critical for microelectronic chip manufacturing for opthalmological application.

Other applications of ArF laser include the production of semiconductor integrated circuits. It also facilitates in situ analysis of mineral samples based on the ablation process.

Krypton Fluoride Laser

The krypton fluoride (KrF) laser is an excimer laser, which emits deep ultraviolet wavelengths. It consists of two elements, krypton and fluorine, as the lasing medium. The principle behind the working of the laser is that krypton absorbs energy from the power source and reacts with fluorine gas to produce krypton fluoride, a temporary complex. This complex is a cable of stimulated emissions, resulting in an exciplex laser in the near ultraviolet region of the electromagnetic spectrum, having a wavelength of 248 nm.

Most excimer lasers are a combination of a noble gas and a halide. They are built on an electronic transition principle, and the emitted radiation is predominantly in the UV spectral range. Commonly used excimer lasers are krypton fluoride, argon fluoride, xenon fluoride, and xenon chloride. Safety precautions need to be exercised while working with this laser, as it damages the human eye when exposed to the UV radiation. Human flesh also needs to be protected from the potentially carcinogenic effects of UV radiation from this laser.

Applications of krypton fluoride laser are mainly in deep-ultraviolet photolithography. A number of microelectronic devices are manufactured by photolithography. After the discovery of excimer lasers in 1982, lamp-based lithography techniques became obsolete. Apart from photolithography, krypton fluoride lasers are used in nuclear fusion energy research, production of soft X-ray emissions, micro machining of plastics, composite materials, and organic tissues. Since the krypton fluoride laser emits deep ultraviolet radiation it is used in medical surgery and therapy.

Laser Properties	
Laser type	Gas
Pump source	High Voltage Electric Discharge
Operating Wavelength	248 nm

Applications

The most important application of a krypton fluoride laser is in photolithography, for the manufacture of microelectronic devices. Use of this laser for lithography has not only replaced the use of mercury-xenon lamps, but has also provided much more precise and finer resolution solutions to photolithography techniques. Some of the other applications of the krypton fluoride laser are given below:

- Nuclear fusion energy research;
- Production of soft X-rays from plasma;
- Micromachining of materials such as plastics, crystals, and composites;
- Medical therapy and surgery;
- Dental surgery.

References

- Gas-lasers: rp-photonics.com, Retrieved 13 June 2018
- He-Ne-laser, microwave-radar, electronics: daenotes.com, Retrieved 23 March 2018
- Argon-laser, microwave-radar, electronics: daenotes.com, Retrieved 30 April 2018
- Laser-types-gas: worldoflasers.com, Retrieved 10 July 2018
- What-is-an-excimer-laser, technology: gigaphoton.com, Retrieved 17 May 2018

5

Semiconductor Lasers and Laser Diodes

A laser diode is a semiconductor laser. It is a semiconductor device in which the laser beam is created at the junction of the diode. Semiconductor lasers are compact-sized and cheap lasers. This chapter discusses in detail semiconductor lasers and laser diodes and includes topics such as quantum dot laser, quantum cascade laser, laser diode, vertical-cavity surface-emitting laser, etc.

Semiconductor Lasers

Semiconductor lasers are lasers based on semiconductor gain media, where optical gain is usually achieved by stimulated emission at an inter band transition under conditions of a high carrier density in the conduction band.

The physical origin of gain in a semiconductor (for the usual case of an inter band transition) is illustrated in figure below without pumping, most of the electrons are in the valence band. A pump beam with a photon energy slightly above the band gap energy can excite electrons into a higher state in the conduction band, from where they quickly decay to states near the bottom of the conduction band. At the same time, the holes generated in the valence band move to the top of the valence band. Electrons in the conduction band can then recombine with these holes, emitting photons with an energy near the band gap energy. This process can also be stimulated by incoming photons with suitable energy. A quantitative description can be based on the Fermi–Dirac distributions for electrons in both bands.

Most semiconductor lasers are laser diodes, which are pumped with an electrical current in a region where an n-doped and a p-doped semiconductor material meet. However, there are also optically pumped semiconductor lasers, where carriers are generated by absorbed pump light, and quantum cascade lasers, where intra band transitions are utilized.

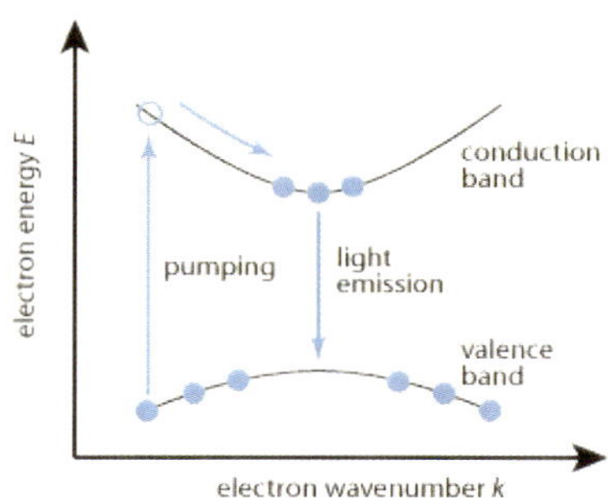

Figure: Physical origin of gain in a semiconductor.

Common materials for semiconductor lasers (and for other optoelectronic devices) are:

- GaAs (gallium arsenide);
- AlGaAs (aluminum gallium arsenide);
- GaP (gallium phosphide);
- InGaP (indium gallium phosphide);
- GaN (gallium nitride);
- InGaAs (indium gallium arsenide);
- GaInNAs (indium gallium arsenide nitride)
- InP (indium phosphide);
- GaInP (gallium indium phosphide).

These are all direct band gap semiconductors; indirect band gap semiconductors such as silicon do not exhibit strong and efficient light emission. As the photon energy of a laser diode is close to the band gap energy, compositions with different band gap energies allow for different emission wavelengths. For the ternary and quaternary semiconductor compounds, the band gap energy can be continuously varied in some substantial range. In AlGaAs = $Al_x Ga_{1-x} As$, for example, an increased aluminum content (increased x) causes an increase in the band gap energy.

While the most common semiconductor lasers are operating in the near-infrared spectral region, some others generate red light (e.g. in GaInP-based laser pointers) or blue or violet light (with gallium nitrides). For mid-infrared emission, there are e.g. lead selenide (PbSe) lasers (lead salt lasers) and quantum cascade lasers.

Apart from the above-mentioned inorganic semiconductors, organic semiconductor compounds might also be used for semiconductor lasers. The corresponding technology is by far not mature, but its development is pursued because of the attractive prospect of finding a way for cheap mass production of such lasers. So far, only optically pumped organic semiconductor lasers have been demonstrated, whereas for various reasons it is difficult to achieve a high efficiency with electrical pumping.

Types of Semiconductor Lasers

There is a great variety of different semiconductor lasers, spanning wide parameter regions and many different application areas:

- Small edge-emitting laser diodes generate a few milliwatts (or up to 0.5 W) of output power in a beam with high beam quality. They are used e.g. in laser pointers, in CD players, and for optical fiber communications.
- External cavity diode lasers contain a laser diode as the gain medium of a longer

laser cavity. They are often wavelength-tunable and exhibit a small emission line width.

- Both monolithic and external-cavity low-power levels can also be mode-locked for ultra-short pulse generation.
- Broad area laser diodes generate up to a few watts of output power, but with significantly poorer beam quality.
- High-power diode bars contain an array of broad-area emitters, generating tens of watts with poor beam quality.
- High-power stacked diode bars contain stacks of diode bars for the generation of extremely high powers of hundreds or thousands of watts.
- Surface-emitting lasers (VCSELs) emit the laser radiation in a direction perpendicular to the wafer, delivering a few milliwatts with high beam quality.
- Optically pumped surface-emitting external-cavity semiconductor lasers (VECSELs) are capable of generating multi-watt output powers with excellent beam quality, even in mode-locked operation.
- Quantum cascade lasers operate on intra band transitions (rather than inter band transitions) and usually emit in the mid-infrared region, sometimes in the terahertz region. They are used e.g. for trace gas analysis.

Basic Structure of Semiconductor Lasers

The basic structure of a semiconductor laser is shown in figure below. The active layer (light emission layer) sandwiched between the p- and n-type clad layers (double hetero structure) is formed on an n-type substrate, and voltage is applied across the p-n junction from the electrodes. Both edges of the active layer has mirror-like surface. When forward voltage is applied, electrons conbine with holes at the p-n junction and emitt the light. This light is not a laser yet; it is confined within the active layer because the refractive index of the clad layers are lower than that of the active layer. In addition, both ends of the active layer act as a reflecting mirror where the light reciprocates in the active layer. Then, the light is amplified by the stimulated emission process and laser oscillation is generated.

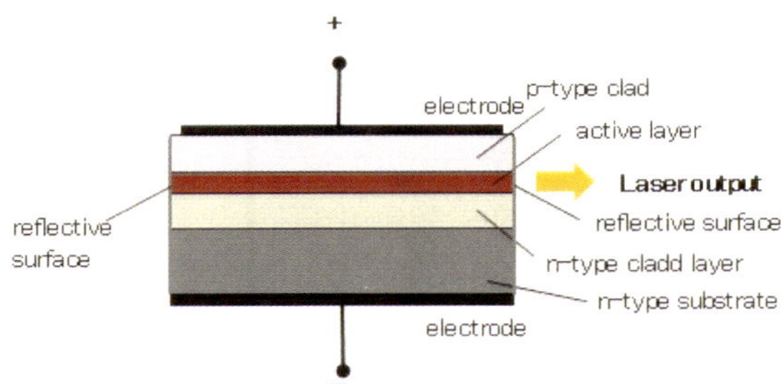

Fig: Basic structure of a semiconductor laser

Typical Characteristics and Applications

Some typical aspects of semiconductor lasers are:

- Electrical pumping with moderate voltages and high efficiency is possible particularly for high-power diode lasers, and allows their use e.g. as pump sources for highly efficient solid-state lasers (→ diode-pumped lasers).
- A wide range of wavelengths are accessible with different devices, covering much of the visible, near-infrared and mid-infrared spectral region. Some devices also allow for wavelength tuning.
- Small laser diodes allow fast switching and modulation of the optical power, allowing their use e.g. in transmitters of fiber-optic links.

Such characteristics have made semiconductor lasers the technologically most important type of lasers. Their applications are extremely widespread, including areas as diverse as optical data transmission, optical data storage, metrology, spectroscopy, material processing, pumping solid-state lasers (→ diode-pumped lasers), and various kinds of medical treatments.

Pulsed Output

Most semiconductor lasers generate a continuous output. Due to their very limited energy storage capability (low upper-state lifetime), they are not very suitable for pulse generation with Q switching, but quasi-continuous-wave operation often allows for significantly enhanced powers. Also, semiconductor lasers can be used for the generation of ultrashort pulses with mode locking or gain switching. The average output powers in short pulses are usually limited to at most a few milliwatts, except for optically pumped surface-emitting external-cavity semiconductor lasers (VECSELs), which can generate multi-watt average output powers in picosecond pulses with multi-gigahertz repetition rates.

Modulation and Stabilization

A particular advantage of the short upper-state lifetime is the capability of semiconductor lasers to be modulated with very high frequencies, which can be tens of gigahertz for VCSELs. This is exploited mainly in optical data transmission, but also in spectroscopy, for the stabilization of lasers to reference cavities, etc.

Application/Uses of Semiconductor Lasers

- The semiconductor laser can be pulsed at varying rate and pulse widths. Therefore this laser is a natural transmitter of digital data.
- Semiconductor laser is well suited for interface with fiber optic cables used in communication.

Quantum Dot Laser

Quantum dots are small semiconductor structures in a host medium with higher electronic (energy) band gap. Dot dimensions typically range from 2 to 10 nm or ~10 to 50 atoms. More specifically, quantum dots for laser diodes are self-organized nanostructures that form spontaneously and controllably on a lattice-mismatched III-V substrate during epitaxial growth processes, like MBE or MOCVD. They function by localizing charge carriers, i.e., electrons and holes (excitons), through quantum confinement. This restricts the three translational degrees of freedom and can dramatically enhance useful electronic properties. At the moment, quantum dots are perhaps the best practical example of emerging nanotechnologies. Industrial applications range from optoelectronics and electronics to medical and biological.

Quantum dots constitute the gain medium in virtually all of Innolume's semiconductor diode lasers and related products. This is because their composition, discreteness, and 3D carrier confinement provide valuable laser performance advantages over conventional quantum well hetero structures that offer only 1D confinement and exhibit strain limits. Benefits include reduced threshold current, temperature independence, broadened gain spectrum, and low relative intensity noise.

A quantum dot laser is a semiconductor laser that uses quantum dots as the active laser medium in its light emitting region.

Development of Quantum Dot Lasers

The laser operation is based on producing radiative emission by coupling electrons and holes at non equilibrium conditions to an optical field. The advantages of quantum well lasers on traditional lasers first predicted in 1970s, and first quantum well lasers which were very inefficient were demonstrated at those dates.

The advantages recognized were:

- The confinement and nature of the electronic density of states result in more efficient devices operating at lower threshold currents than lasers with bulk active layers. The laser threshold current density can be reduced by decreasing the thickness of the active layer.
- Discrete energy levels provide a means of "tuning" the resulting wavelength of the material. Since the thickness of the quantum well-depends on the desired spacing between energy levels, tuning can be done by changing the quantum well dimensions or thickness. For energy levels of greater than a few tens of meV's, the critical dimension is approximately a few hundred angstroms.

The inefficiency of quantum well lasers were eliminated in 1980s by the use of new materials growth capabilities (molecular beam epitaxy), and optimization of the hetero structure laser design.

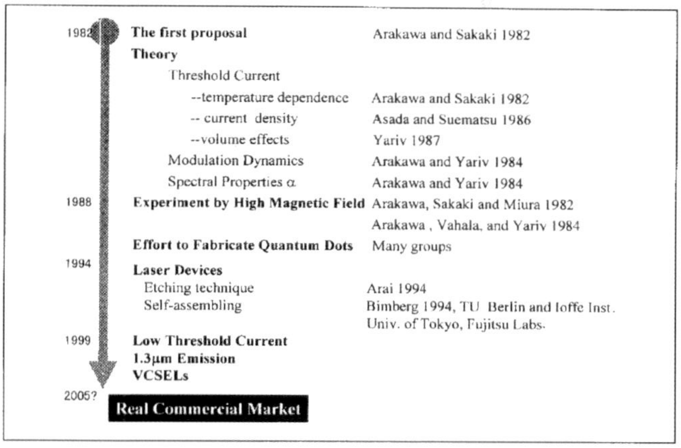

Figure: The historical evolution of QD lasers from the beginning

Since the quantum confinement in a QD is in all three dimensions, tunability of a quantum dot laser (QDL) is higher than a quantum well laser (QWL). The concept of semiconductor QDs was proposed for semiconductor laser applications by Arakawa and Sakaki in 1982, predicting suppression of temperature dependence of the threshold current. Henceforth, reduction in threshold current density, reduction in total threshold current, enhanced differential gain and high spectral purity/no-chirping were theoretically discussed.

At this point, we should examine the basics of laser operation. A laser utilizes stimulated that is triggered by an incident photon of the same energy. This occurs when a medium has more population of electrons in the excited quantum level than in the ground level. This artificially situation, called population inversion, is produced by either electrical stimulation (electroluminescence) or optical stimulation and is different from the spontaneous emission, whereby the electron returns to the ground state in the natural course (within the lifetime of the excited states), even in the absence of any photon to stimulate it.

In figure above, stimulated recombination of electron-hole pairs takes place in the GaAs quantum well region, where the confinement of carriers and the confinement of the optical mode enhance the interaction between carriers and radiation.

In figure below, we can observe the changes in density of states for different dimensionalities. The population inversion necessary for lasing occurs more efficiently as the active layer material is scaled down from bulk (3D) to QDs (0D). However, the advantages in operation depends both on the absolute size of the nanostructures in the active region, and on the uniformity of size. A broad distribution of sizes smears the density of states, producing behavior similar to that of bulk material.

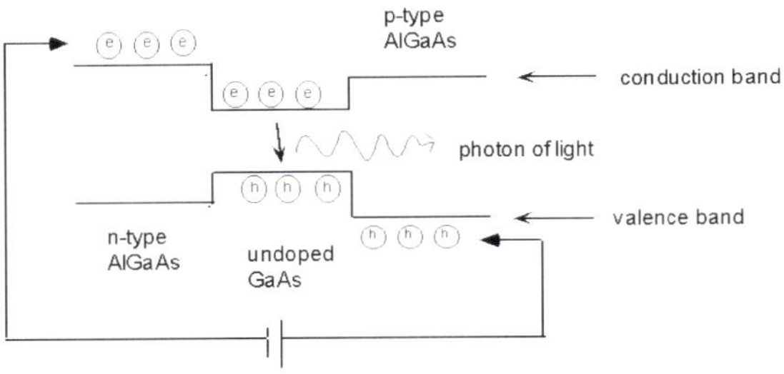

Figure: Schematic of a semiconductor laser

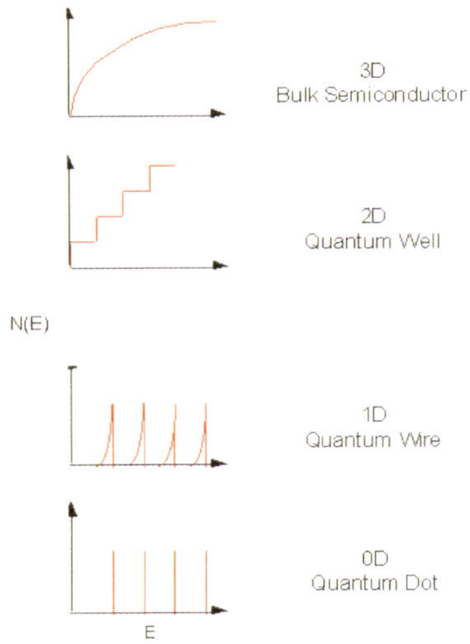

Figure: Electronic states as a function of structure size

QD lasers acquired more importance after significant progress in nanostructure growth in the 1990's such as the self-assembling growth technique for InAs QDs. The first demonstration of a quantum dot laser with high threshold density was reported by Ledentsov and colleagues in 1994. Bimberg et al. (1996) achieved improved operation by increasing the density of the QD structures, stacking successive, strain-aligned rows of QDs and therefore achieving vertical as well as lateral coupling of the QDs. In addition to utilizing their quantum size effects in edge emitting lasers, self-assembled QDs have also been incorporated within vertical cavity surface emitting lasers.

QD lasers are not as temperature dependent as traditional semiconductor lasers. This theory was utilized by applications and in 2004; temperature-independent QD lasers were invented in Fujitsu Laboratories.

We can summarize the predicted advantages of QD-lasers as:

1. Emits light at wavelengths determined by the energy levels of the dots, rather than the band gap energy. Thus, they offer the possibility of improved device performance and increased flexibility to adjust the wavelength.

2. Has the maximum material gain and differential gain, at least 2-3 orders higher than QW lasers.
3. Advantages of small volume:
   a. Low power high frequency operation,
   b. Large modulation bandwidth,
   c. Small dynamic chirp,
   d. Small line width enhancement factor,
   e. Low threshold current.
4. Shows superior temperature stability of the threshold current. The threshold current is given by the relation:

   $$I_{\text{üüüüüüüüthth}}(T) = I \qquad (T \quad).\exp\left(\left(T-(T \quad)\right)/(T \quad)\right),$$

   Where, T is the active region temperature, (T_{ref}) is the reference temperature, and (T_0) is an empirically-determined "characteristic temperature", which is itself a function of temperature and device length. In QDLs To can be high, because one can effectively decouple electron phonon interaction by increasing the inter sub band separation. This leads to undiminished room-temperature performance without external thermal stabilization.
5. QD lasers suppress the diffusion of non-equilibrium carriers, resulting in reduced leakage from the active region.

Basic Characteristics of QD Lasers

In a laser, the stimulated emission is amplified by passing the emitted photons to stimulate emission at other locations. The basic components of a laser are:

- An active medium (gain medium which is the QD in our case) where population inversion is created by a proper pumping mechanism. The spontaneously emitted photons at some site in the medium stimulate emission at other sites as it travels through it.
- An energy pump source (electric power supply for QDLs).
- Two reflectors (rear mirror and output coupler) to reflect the light in phase (determined by the length of the cavity) so that the light will be further amplified by the active medium in each round-trip (multi pass amplification). The output is partially transmitted through a partially trans missive output coupler where the output exits as a laser beam.

Figure below shows schematic view of the band structure of a typical quantum dot laser. An ideal QD laser consists of a 3D-array of dots with equal size and shape (middle of the figure), surrounded by a higher band-gap material which confines the injected carriers. The whole structure is embedded in an optical waveguide consisting of lower and upper cladding layers (n-doped and pdoped shields).

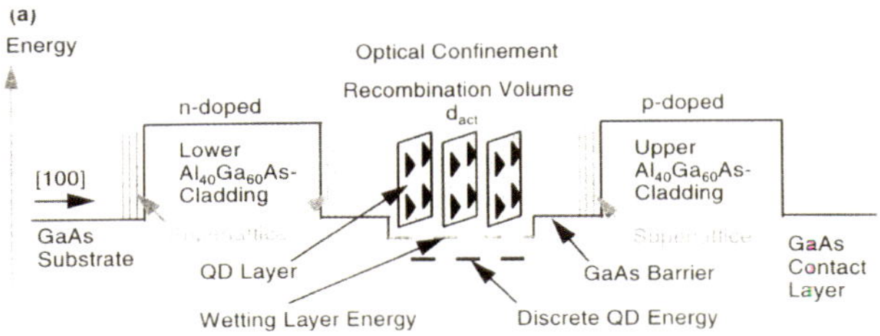

Figure— Band structure of a quantum dot laser with self-organized dots under forward bias

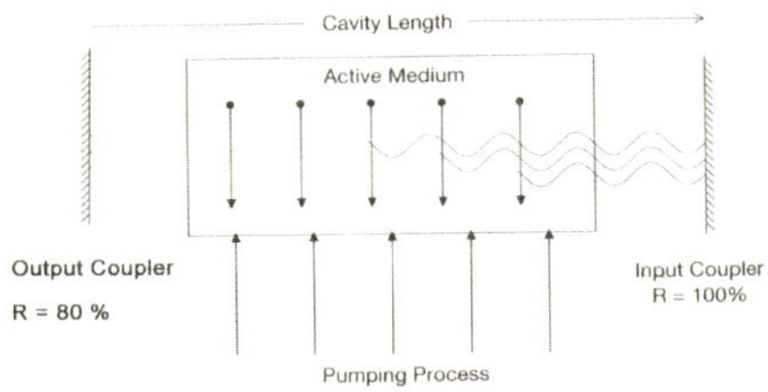

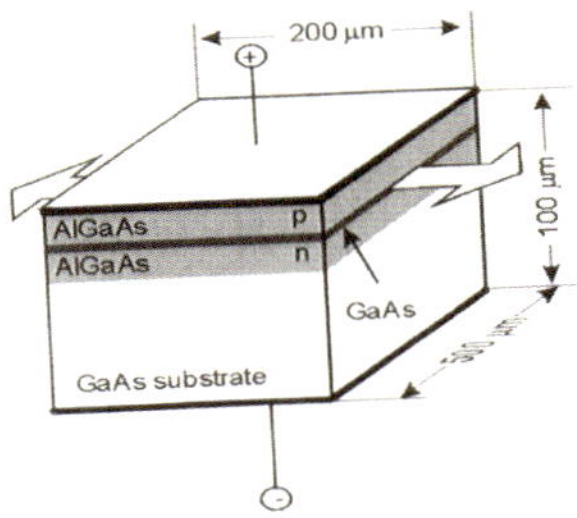

QD lasers are established by spontaneous formation of QDs at growth temperatures between 460 and 550°C. Fabrication method for such QDs is Stranski-Krastanov growth. The low growth temperature and the low dot density can cause several problems concerning threshold and gain. The cladding layer and the GaAs QD barrier typically grown at these lower temperatures are a possible source for current leakage and non-radiative recombination. On the other hand, the QDs exhibit some intermixing with the surrounding barrier material if temperatures of about 700°C are used to grow high quality cladding layers.

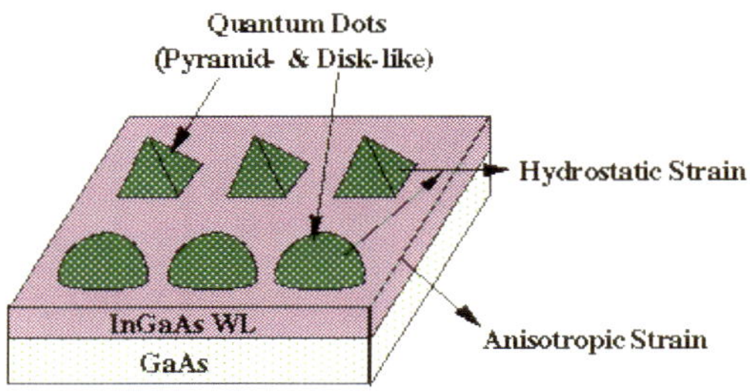

The self-organization of nanoscale three-dimensional coherent strained islands following Stranski-Krastanov growth mechanism is considered as the most promising way of in-situ QDs fabrication. The ordered arrays so formed may result in distributed feedback and in stabilization of single-mode lasing. In addition, intrinsically buried QDs spatially localize carriers and prevent them from recombining non-radiatively at resonator facets. Overheating of facets at high power operation may thus be avoided. A real challenge lies in the optimization of growth parameters to achieve a dense and uniform array of QDs, identical in size and shape.

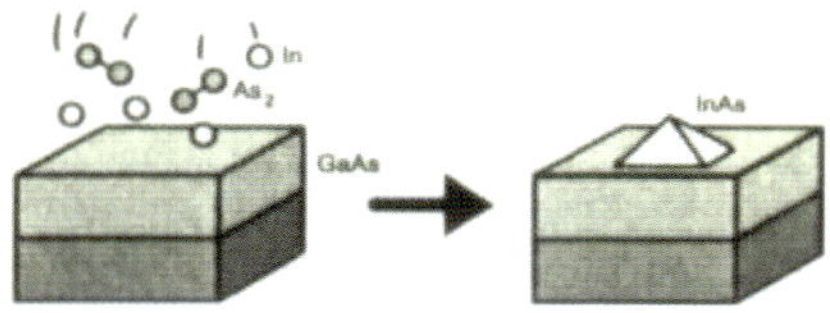

Figure: Representation of different approaches to fabrication of nanostructures: self-organized growth of nanostructures

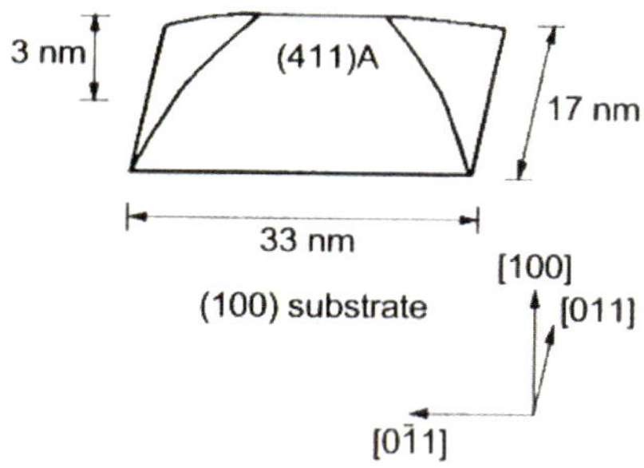

Figure: Estimated shape of a self-assembled QD made of InGaAs

Required Characteristics for Quantum Dot Laser Applications

Quantum dot lasers utilize an oscillator strength that is condensed into a narrow energy width. Because of that reason, the absolute energy level of the QDs should be the same. In other words, the size, shape and alloy composition of QDs should be close to identical. Therefore, the inhomogeneous broadening of QD luminescence is eliminated, and real concentration of the electron energy states can be obtained. If a macroscopic physical parameter is desired, such as light output in laser devices, the density (the number of interacting QDs) should be as high as possible. The reduction of non radiative centers in QDs is important for QDL applications. Nanostructures made by high-energy beam patterning cannot be used damage is incurred from the beam around the nanostructures. Since the surface-to-volume ratio of QDs is drastically increased compared to QWs, this type of damage around the surface of self-assembled QDs is critical for the development of the QDL applications.

QDs are put into layered structures to create lasers. At this point, electrical control is very important because an electric field applied to the structure can change certain physical properties of QDs in a desirable way and carriers can be injected into the structure to create light emission.

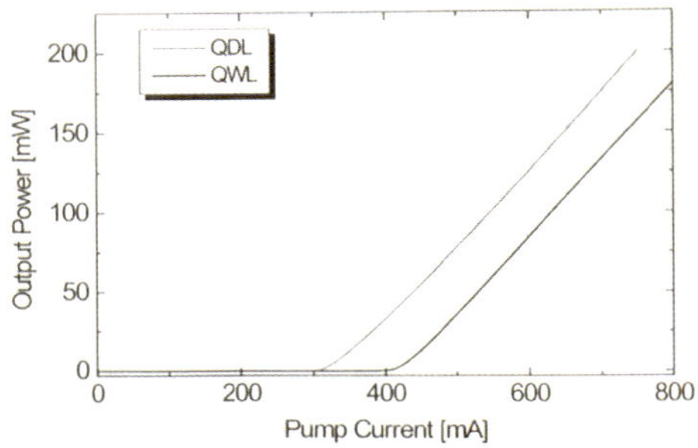

In order for QD lasers compete with QW lasers, two major issues have to be addressed:

- A large array of QDs has to be used because their active volume is very small. An array of QDs with a narrow size distribution has to be produced to reduce in homogeneous broadening. Furthermore, that array has to be without defects that degrade the optical emission by providing alternate non radiative defect channels.
- The phonon bottleneck created by confinement limits the number of states that are efficiently coupled by phonons due to energy conversation. Therefore, it also limits the relaxation of excited carriers into lasing states. This bottleneck causes degradation of stimulated emission. However, other mechanisms can be used to suppress that bottleneck effect.

Different Types of Quantum Dot Lasers

High Speed Quantum Dot Lasers

There are several epitaxials were proposed to get the predicted advantages of QD lasers, among them are: overgrowth of QDs with quantum well layers, stacking of quantum dots, close stacking of quantum dots leading to the vertical coupling of quantum dot layers, p-doping of the GaAs barrier layers.

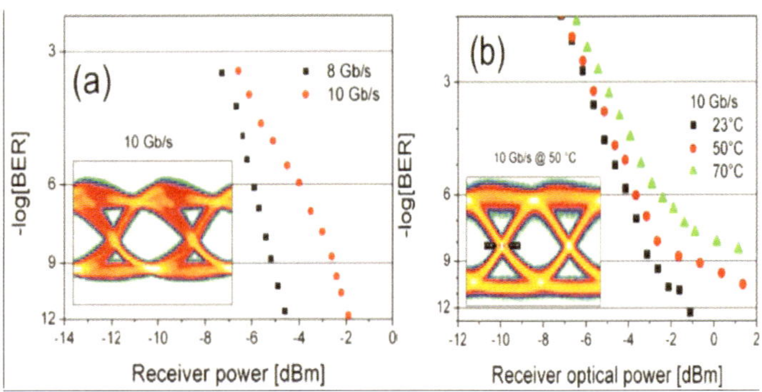

Figure: BER measurment of QD laser module at 8 Gb/s and 10 Gb/s (a) and at 10 Gb/s for diffrent temperatures (b) inset shows the coresponding eye patterns

- Directly modulated quantum dot lasers:

  Being the key point of the fiber-based data com application, directly modulated quantum dot lasers could convert electrical signals into digital optical signals at the rate of around 10Gb/s. The modulation speed needs to be further improved, the power consumption should be reduced and the temperature performance needs to be better.

Mode-Locked Quantum Dot Lasers

- With the applications of Mode-Locked quantum dot lasers, several advantages could be received: short optical pulses, narrow spectral width with a small footprint device. Besides, Mode-Locked quantum dot lasers are able to provide a much broader gain spectrum (>50nm), longer cavities (approximately 1cm), sub-ps width and low α factor which leads to low chirp.

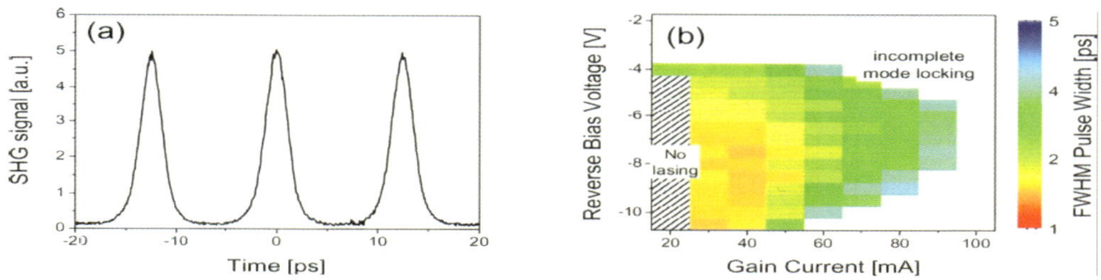

InP Based Quantum Dot Lasers

- Compared with QW lasers, the emission wavelength of the InP based quantum dot lasers is much lower (0.2nm/K compared to 0.55nm/K). This property could allow this kind of quantum dot lasers operate within a much wider temperature range. Although there still exist some limitations in speed due to the inhomogeneous line width broadening, the data transmission could still be possibly over 10Gb/s for InP based quantum dot lasers.

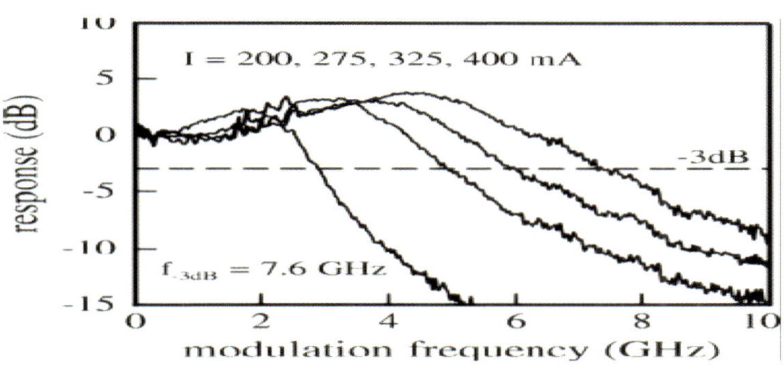

Figure– Small signal modulation response for a Q Dash DFB laser at four different pulsed (10 μs pulses at a duty cycle of 2%) bias levels

High Power Quantum Dot Lasers

With several promising properties of the quantum dot materials, it is widely realized that quantum dot lasers are able to get a good power performance. The advantages of quantum dot materials to be suitable applied to high power application fields are: zero line width enhancement factor, the free geometric parameters of the quantum dots, e. g. quantum dots size, dots density and size distribution could allow to get the gain without considering the material composition. Extra expensive cooling by Peltier elements is then not needed.

QD Lasers for Cooler Less Pump Sources

The devices with these properties are recently developed based on the GaInAs/Ga(Al)As QD layers emitting at 920nm. In this research, the size of the quantum dot is reduced by modifying the growth parameter and In composition with a constant emission wavelength of the transition. A power splitting of 65ev could be received at room temperature wavelength of 920nm with the size reduced quantum dot structures.

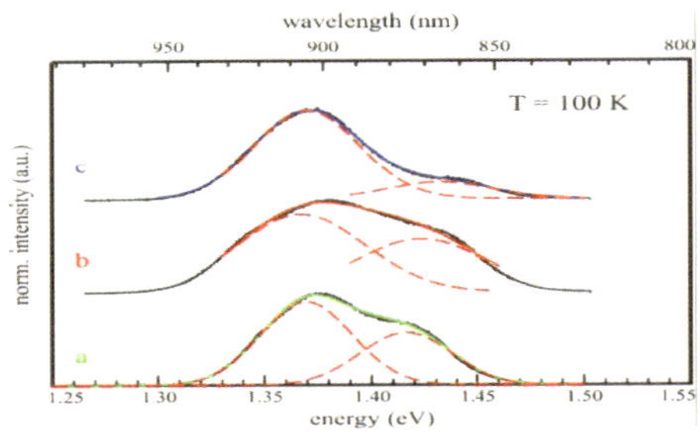

Figure – PL spectra of dot layers with different dot sizes resulting in different transition energy splittings between fundamental and first excited state transitions

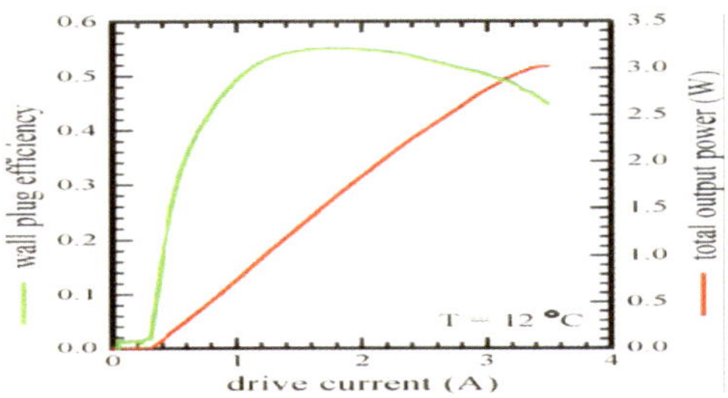

Figure: Total output power and wall-plug efficiency of a quantum dot high power laser with 1 mm cavity length and 100 μm broad contact stripes

Single Mode Tapered Lasers

New device geometry is used to get a similar performance as multi-mode emitting devices. Such kind of lasers enables the amplification of the single mode during the propagation. Compared with quantum well lasers, the wavelength shift is smaller due to the better temperature performance (temperature sensitivity).

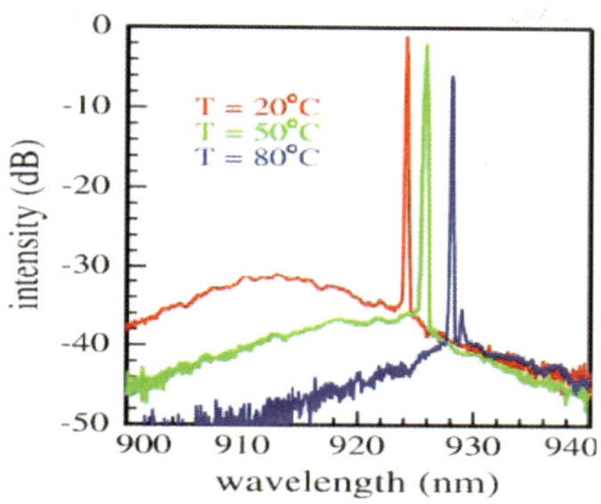

Figure – Emission spectra for 3 different temperatures of a single mode emitting QD tapered laser

Quantum Cascade Laser

Quantum Cascade Lasers (QCL) are semiconductor lasers that emit in the mid- and long-wave IR bands, and are finding new applications in precision sensing, spectroscopy, medical, and military applications. Their wide tuning range and fast response time allow for faster and more precise compact trace element detectors and gas analyzers that are replacing slower and larger FTIR, mass spectroscopy, and photo thermal micro spectroscopy systems.

Since the first operational QCL emitted light in 1994 tremendous effort has been put into making them more robust, versatile, and manufacturable. Quantum cascade lasers are used in industrial exhaust sensing, safe-distance explosives detectors, light sources for infrared imaging systems, medical diagnostic breath analyzers, and a variety of other sensing and spectroscopic applications. The number of markets for QCLs is growing at an ever-increasing rate as scientists and manufacturers gain more experience with them.

QCLs Operate in a Fundamentally Different Way Than Diode Lasers

- Diode lasers depend on the process of electron-hole recombination: an electron from the conduction band recombines with a hole in the valence band, and in the process a single photon is emitted.
- Diode lasers are limited to about 2.5 μm wave length because the wavelength is determined by the recombination energy, or band gap, of the material system used to fabricate the device. Different material combinations result in different band gaps, but there is a limit to the materials that can be used to make a diode laser.
- This figure illustrates the principle on which QCLs operate:

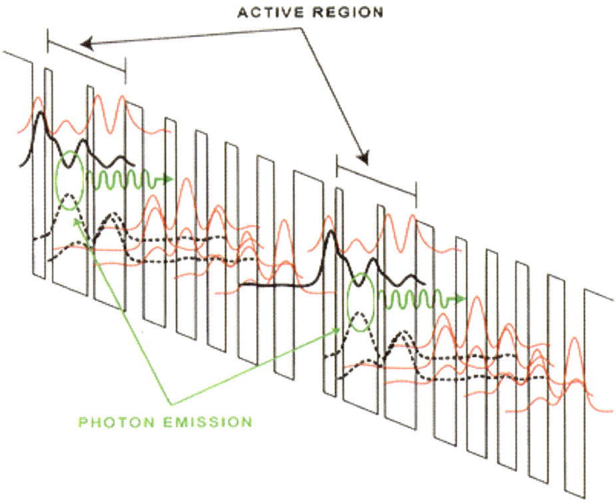

Figure: Quantum cascade laser energy band

- Quantum cascade lasers are comprised of dozens of alternating layers of semiconductor material, forming quantum energy wells that confine the electrons to particular energy states. As each electron traverses the lasing medium it transitions from one quantum well to the next, driven by the voltage applied across the device. At precisely engineered locations, called the "active region," the electron transitions from one valence band energy state to a lower one and in the process emits a photon. The electron continues through the structure and when it encounters the next active region it transitions again and emits another photon. The QCL may have as many as 75 active regions, and each electron generates that many photons as it traverses the structure.

The output wavelength is determined by the structure of the layers rather than the lasing material, and that means device fabricators can tailor the wavelength in a way that can't be achieved with diode lasers. While diode laser output wavelength is limited to ~2.5 μm, QCLs operate at much longer wavelengths: mid-wave infrared production devices up to 11 μm are available, and some 25 μm emitters have been made on an experimental basis. This transmission range is useful because of the large number of absorption lines exhibited by common target gases that fall within this band.

Terahertz QCLs are now being commercialized– some emitting in the range of 100 μm to 150 μm. More complex gas molecules absorb at these longer wavelengths, and greater measurement accuracy is possible with QCL-based systems than with current technologies. QCLs allow for remote sensing of these complex molecules, whereas most traditional methods require a sample of the subject material. Some technical challenges remain before these systems are commercially viable on a large scale, but the technology is advancing rapidly.

Applications

A number of applications have been well served by mid-IR quantum cascade lasers. This region of the spectrum is interesting because of a combination of two facts. The atmosphere is (at least somewhat) transparent at these wavelengths and many species of interest have strong fundamental absorptions that make it possible to detect and identify them. Figure below is a depiction of the transmission of the atmosphere in the visible through the mid-infrared portion of the electromagnetic spectrum. Significant regions of high transmission exist in the mid-IR and can be exploited to transmit powers at these wavelengths for various applications. Free-space communication, infrared countermeasures, remote imaging, beaconing, and illumination are all enabled by QCLs at an appropriate wavelength. Figure below is a graph of the mid-infrared portion of the spectrum with a number of species placed where their strong absorptions occur. The mid-IR is rich in information for those wishing to probe, detect, image, or quantify these and many other species including explosives, nerve agents, and toxins. Detection limits in the parts per trillion range and discrimination between similar species are possible.

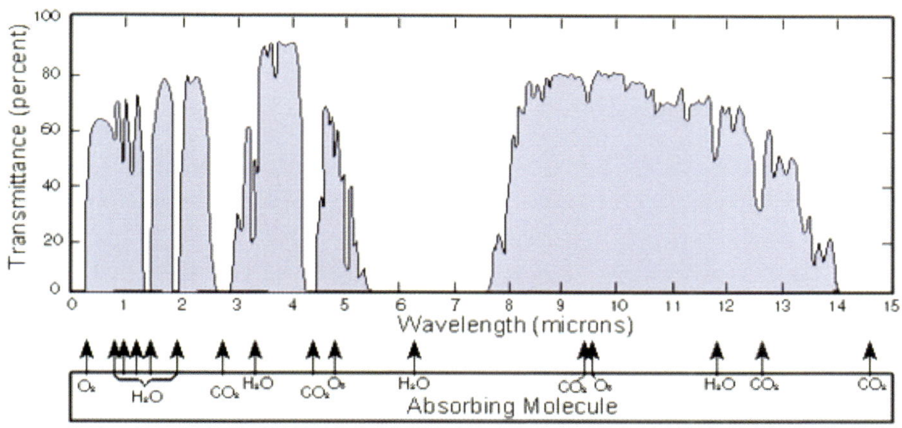

Figure: Transmission of light through the atmosphere *from the visible to the mid-infrared*

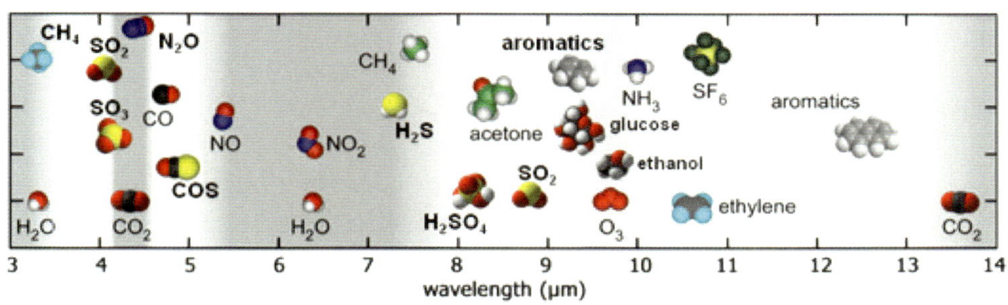

Figure: Graphical representation of the location of strong absorptions of molecules of interest

A key application for QCLs is stand-off explosives detection. In this developing field researchers have set the ambitious goal of detecting and discriminating nanogram quantities of various explosives at distances up to 50 m with eye-safe lasers. There are a number of tactics being employed, one approach being thermal imaging. When a compound absorbs infrared light, it re-emits most of the absorbed light isotropically as heat which can be imaged by infrared cameras. Since each analysis has a unique absorption spectrum, each will heat selectively as the midir source is tuned through these absorptions and may be identified unambiguously by analysis of the multi-spectral or hyper spectral data cube produced.

While quantum cascade lasers serve as the engines for new techniques in spectroscopy in the mid-IR, they also can provide raw power at new performance levels. Powers exceeding 5 W have been demonstrated from single room-temperature devices. Combining performance such as this with ruggedized packaging has enabled a new generation of Infrared Countermeasure (IRCM) devices. High-power, solid-state lasers that operate in mid-infrared "atmospheric windows" can be used by pointer-trackers to disable the heat seeking mechanism employed on surface-to-air missiles, thus safeguarding soldiers in battlefield situations. Multiple "socket" architectures where a set of QCLs have been made to be collinear have been productized in militarily hardened packages. These units can produce in excess of 15 W and have completed a series of stringent environmental tests, including helicopter flight testing.

Quantum cascade lasers have matured from scientific lab experiments through a technological acceptance phase and now have entered a technological dependence phase as devices have been deployed in real-world applications such as gas and explosive detection, imaging, and countermeasures that can only be serviced by this technology. The optical physics of these devices is unique and it promises to provide continuing improvements in the performance.

Vertical-cavity Surface-emitting Laser

Vertical-cavity surface-emitting lasers (VCSELs) are a relatively recent type of semiconductor lasers. VCSELs were first invented in the mid-1980's. Very soon, VCSELs gained a reputation as a superior technology for short reach applications such as fiber channel, Ethernet and intra-systems links. Then, within the first two years of commercial availability (1996), VCSELs became the technology of choice for short range datacom and local area networks, effectively displacing edge-emitter lasers. This success was mainly due to the VCSEL's lower manufacturing costs and higher reliability compared to edge-emitters.

Princeton optronics has developed the key technologies resulting in the world's highest power single VCSEL devices and 2-D arrays. We have successfully demonstrated single devices with >5W CW output power and large 2D arrays with >230W CW output power. We have made single mode devices of 1W output power and single mode arrays with power of >100W which are coupled to 100u, 0.22NA fiber. The highest wall plug efficiency of these devices and arrays is 56%. We have made arrays which deliver 1kW/cm^2 in CW operation and 4.2 kW/cm^2 in QCW operation. Princeton Optronics was a participant in the DARPA-SHEDS program, whose main objective was to improve laser diode conversion efficiency.

The VCSEL Structure

Semiconductor lasers consist of layers of semiconductor material grown on top of each other on a substrate (the "epi"). For VCSELs and edge-emitters, this growth is typically done in a molecular-beam-epitaxy (MBE) or metal organic-chemical-vapor-deposition (MOCVD) growth reactor. The grown wafer is then processed accordingly to produce individual devices. Figure below summarizes the differences between VCSEL and edge-emitter processing.

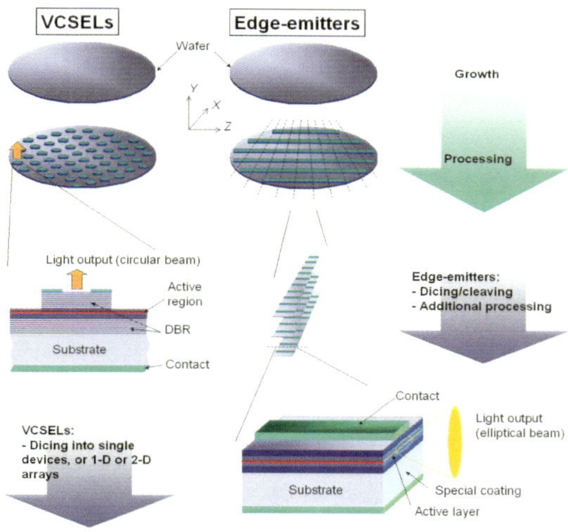

Figure: Comparison of the growth/processing flow of VCSEL and edge-emitter semiconductor lasers

In a VCSEL, the active layer is sandwiched between two highly reflective mirrors (dubbed distributed Bragg reflectors, or DBRs) made up of several quarter-wavelength-thick layers of semiconductors of alternating high and low refractive index. The reflectivity of these mirrors is typically in the range 99.5~99.9%. As a result, the light oscillates perpendicular to the layers and escapes through the top (or bottom) of the device. Current and/or optical confinement is typically achieved through either selective-oxidation of an Aluminum-rich layer, ion-implantation, or even both for certain applications. The VCSELs can be designed for "top-emission" (at the epi/air interface) or "bottom-emission" (through the transparent substrate) in cases where "junction-down" soldering is required for more efficient heat-sinking for example. Figure illustrates different common types of VCSEL structures.

In contrast, edge-emitters are made up of cleaved bars diced from the wafers. Because of the high index of refraction contrast between air and the semiconductor material, the two cleaved facets act as mirrors. Hence, in the case of an edge-emitter, the light oscillates parallel to the layers and escapes side-ways. This simple structural difference between the VCSEL and the edge-emitter has important implications.

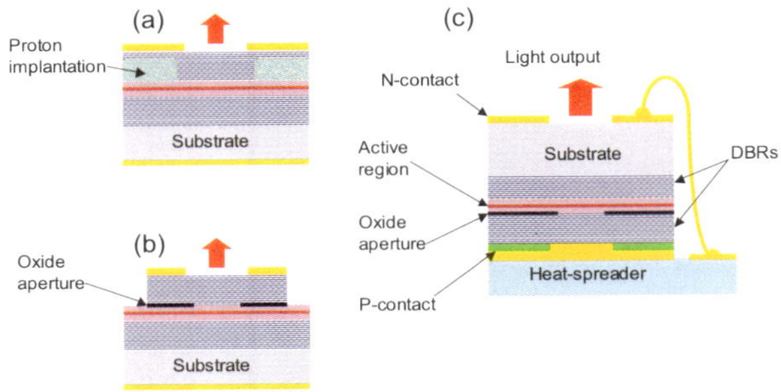

Figure: Three common types of VCSEL structures: (a) a top-emitting structure with proton implantation to confine the current, (b) a selectively-oxidized top-emitting structure to confine the optical modes and/or the current, and (c) a mounted bottom-emitting selectively-oxidized structure

Since VCSELs are grown, processed and tested while still in the wafer form, there is significant economy of scale resulting from the ability to conduct parallel device processing, whereby equipment utilization and yields are maximized and set up times and labor content are minimized. In the case of a VCSEL, the mirrors and active region are sequentially stacked along the Y axis during epitaxial growth. The VCSEL wafer then goes through etching and metalization steps to form the electrical contacts. At this point the wafer goes to test where individual laser devices are characterized on a pass-fail basis. Finally, the wafer is diced and the lasers are binned for either higher-level assembly (typically >95%) or scrap (typically 2W output power) packaged on a high-thermal conductivity sub mount.

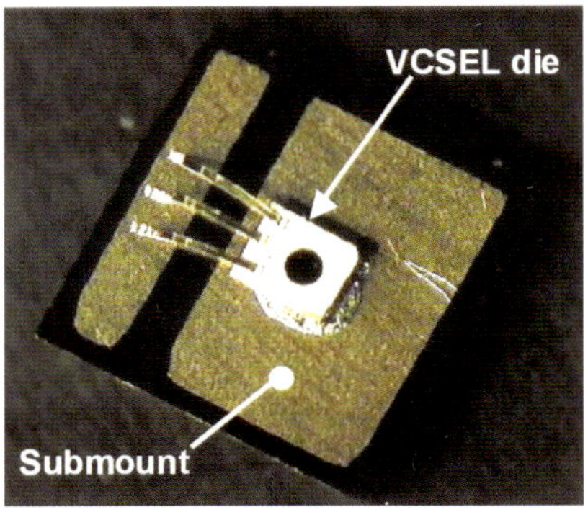

Figure: Packaged high-power VCSEL device (>2W). The sub mount is 2mm x 2mm

In a simple Fabry-Pérot edge-emitter the growth process also occurs along the Y axis, but only to create the active region as mirror coatings are later applied along the Z axis. After epitaxial growth, the wafer goes through the metallization step and is subsequently cleaved along the X axis, forming a series of wafer strips. The wafer strips are then stacked and mounted into a coating fixture. The Z axis edges of the wafer strips are then coated to form the device mirrors. This coating is a critical processing step for edge-emitters, as any coating imperfection will result in early and catastrophic failure of the devices due to catastrophic-optical-damage (COD). After this coating step, the wafer strips are diced to form discrete laser chips, which are then mounted onto carriers. Finally, the laser devices go into test.

It is also important to understand that VCSELs consume less material: in the case of a 3" wafer, a laser manufacturer can build about 15,000 VCSEL devices or approximately 4,000 edge-emitters of similar power levels.

In addition to these advantages, VCSEL also demonstrate excellent dynamic performances such as low threshold currents (a few micro-amps), low noise operation and high-speed digital modulation (10 Gb/s). Furthermore, although VCSELs have been confined to low-power applications – a few milli-Watts at most – they have the inherent potential of producing very high powers by processing large 2-D arrays. In contrast, edge-emitters cannot be processed in 2-D arrays.

VCSEL Advantages

The many advantages offered by the VCSEL technology can be summarized in the following points:

1. Wavelength stability: The lasing wavelength in a VCSEL is very stable, since it is fixed by the short (1~1.5- wavelength thick) Fabry-Perot cavity. Contrary to edge-emitters, VCSELs can only operate in a single longitudinal mode.
2. Wavelength uniformity and spectral width: Growth technology has improved such that VCSEL 3" wafers are produced with less then a 2nm standard deviation for the cavity wavelength. This allows for the fabrication of VCSEL 2-D arrays with little wavelength variation between the elements of the array (<1nm full-width half- maximum spectral width). By contrast, edge-emitter bar-stacks suffer from significant wavelength variations from bar to bar since there is no intrinsic mechanism to stabilize the wavelength, resulting in a wide spectral width (3~5nm FWHM).
3. Temperature sensitivity of wavelength: The emission wavelength in VCSELs is ~5 times less sensitive to temperature variations than in edge-emitters. The reason is that in VCSELs, the lasing wavelength is defined by the optical thickness of the single-longitudinal-mode-cavity and that the temperature dependence of this optical thickness is minimal (the refractive index and physical thickness of the cavity have a weak dependence on temperature). On the other hand, the lasing wavelength in edge-emitters is defined by the peak-gain wavelength, which has a much stronger dependence on temperature. As a consequence, the spectral line- width for high-power arrays (where heating and temperature

gradients can be significant) is much narrower in VCSEL arrays than in edge-emitter-arrays (bar-stacks). Also, over a 20°C change in temperature, the emission wavelength in a VCSEL will vary by less than 1.4nm (compared to ~7nm for edge-emitters).

4. High Temperature operation (Chiller less operation for pumps): VCSEL devices can be operated without refrigeration- because they can be operated at temperatures to 80 deg C, The cooling system becomes very small, rugged and portable with this approach.

5. Higher power per unit area: Edge emitters deliver a maximum of about 500W/ cm^2 because of gap between bar to bar which has to be maintained for coolant flow, while VCSELs are delivering ~1200W/ cm^2 now and can deliver 2-4kW/ cm^2 in near future.

6. Beam quality: VCSELs emit a circular beam. Through proper cavity design VCSELs can also emit in a single transverse mode (circular Gaussian). This simple beam structure greatly reduces the complexity and cost of coupling/ beam-shaping optics (compared to edge-emitters) and increases the coupling efficiency to the fiber or pumped medium. This has been a key selling point for the VCSEL technology in low-power markets.

7. Reliability: Because VCSELs are not subject to catastrophic optical damage (COD), their reliability is much higher than for edge-emitters. Typical FIT values (failures in one billion device-hours) for VCSELs are <10.

8. Manufacturability and yield: Manufacturability of VCSELs has been a key selling point for this technology. Because of complex manufacturing processes and reliability issue related to COD (catastrophic optical damage), edge-emitters have a low yield (edge-emitter 980nm pump chip manufacturers typically only get ~500 chips out of a 2" wafer). On the other hand, yields for VCSELs exceed 90% (corresponds to ~5000 high-power chips from a 2" wafer). In fact, because of its planar attributes, VCSEL manufacturing is identical to standard IC Silicon processing.

9. Scalability: For high-power applications, a key advantage of VCSELs is that they can be directly processed into monolithic 2-D arrays, whereas this is not possible for edge-emitters (only 1-D monolithic arrays are possible). In addition, a complex and thermally inefficient mounting scheme is required to mount edge-emitter bars in stacks.

10. Packaging and heat-sinking: Mounting of large high-power VCSEL 2-D arrays in a "junction-down" configuration is straightforward (similar to micro-processor packaging), making the heat-removal process very efficient, as the heat has to traverse only a few microns of AlGaAs material. Record thermal impedances of <0.16K/W have been demonstrated for 5mm x 5mm 2-D VCSEL arrays.

11. Cost: With the simple processing and heat-sinking technology it becomes much easier to package 2-D VCSEL arrays than an equivalent edge-emitter bar-stack. The established existing silicon industry heat-sinking technology can be used for heat removal for very high power arrays. This will significantly reduce the cost of the high-power module. Currently, cost of the laser bars is the dominant cost for the DPSS lasers.

Vertical-external-cavity Surface-emitting-laser

A vertical-external-cavity surface-emitting laser (VECSEL) is a semiconductor Laser similar to a VCSEL, but the laser resonator is completed with an external mirror and may contain additional optical elements.

Vertical-external-cavity surface-emitting lasers (VECSELs) are becoming important devices for high power generation, intracavity frequency conversion, gain-switching and model locking applications. The external-cavity offers interesting features for mode-locking: good beam quality, the possibility of controlling the transverse laser mode and multi-GHz repetition rates with low tendency to Qs witching. The vertical-external-cavity concept has been applied in optically pumped devices and more recently, to electrically-driven devices. Optically pumped VECSELs are suited to high power applications because of the scalable output power with the size of the fundamental transverse mode. Optically-pumped VECSELs can be passively mode-locked by semiconductor saturable absorber mirrors (SESAM). They have generated picosecond pulses with output powers larger than 200 mW and sub-picosecond pulses when operating in the quasi-soliton mode-locking regime.

The VECSELs have been well developed, in which strong transverse mode control can be provided by optical cavity elements external to the laser chip, which assure that fundamental transverse mode of the laser cavity, the desired operating laser mode, has diameter approximately equal to the gain region diameter. Basic structure of VECSEL devices is depicted in figure below. to illustrate the functions of the various layers. In a VECSEL device, a thin active semiconductor chip as the key element with typical diameters range between 50 and 500 μm, containing both a gain region and multilayer high-reflectivity mirrors, is placed on a heat sink and is excited by an incident optical pump beam. Laser cavity consists of the on chip mirror and an external spherical mirror, which defines the laser transverse mode and also serves as the output coupler. For optically pumped operation, incident pump photons with higher photon energy are absorbed in separate pump-absorbing layers, which also serve as the quantum well barriers, to generate electrons and holes. The excited carriers then diffuse to the smaller bandgap quantum wells or QD gain layers that provide gain to the optical wave, emitting lasing photons with lower photon energy. These separate pump absorption and quantum well or QD lasing emission gain layers facilitate independent optimization of the pump absorption and laser gain properties. Typically, gain layer region thickness covers several periods of this laser mode standing wave.

The pump–laser photon energy difference between the incident pump photons and the emitted laser photons, due to quantum defect, together with contributions from other lasing inefficiencies, has to be dissipated as heat from the device active region. Heat dissipation from the VECSEL active semiconductor chip is provided by heat spreaders connected to heat sinks: either a soldered heat spreader below the mirror structure or a transparent heat spreader above the surface window of the chip, or possibly both. It should be pointed out that the issues of good heat dissipation and heat sinking are critical for high-power operation of all semiconductor lasers. Without these, temperature of the active region would rise and excited carriers would escape thermally from the quantum wells or QDs gain layer into the barrier region, thus depleting laser gain and turning the laser off in a thermal rollover process. Such a thermal dominant rollover mechanism typically limits the output power of VECSEL devices. Fortunately, the configuration of VECSELs implements good heat sinking by placement of the transparent intracavity heat spreaders in direct contact with the laser gain region without thermally resistive laser mirrors in the path of heat dissipation, and have the inherent potential of producing very high powers by processing large 2-D arrays.

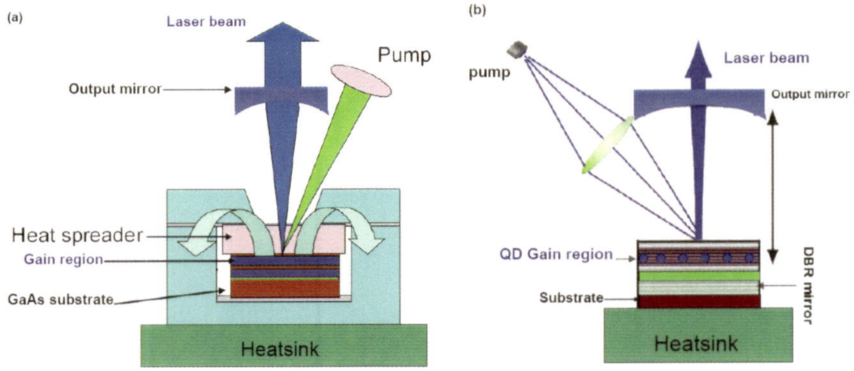

Fig: Structure configuration of VECSEL device (a) and QD VECSEL device (b)

Basic configuration of VECSEL lasers enables their many key advantageous properties, which can be summarized in the following aspects:

- Wide wavelength coverage, wavelength stability and less temperature sensitivity of wavelength: Using numerous compound semiconductor materials for different material compositions with different band gap energies enables different lasing emission wavelengths. The use of dimensionally, or quantum-confined semiconductor active regions allows further control of the VECSEL emission wavelengths. By adjusting the thickness of quantum wells or diameter of quantum wires and QDs, as well as the composition of confining barrier layers, the quantum-confined electron and hole energy levels are shifted and VECSEL acquires the additional fine control of laser emission wavelength. Moreover, the lasing wavelength in a VECSEL is very stable. Note that the emission wavelength in VECSELs is ~5 times less sensitive to temperature variations than in edge-emitting lasers. The reason is that in VECSELs, the lasing wavelength is defined by the optical thickness of the single longitudinal- mode-cavity and that the temperature dependence of this optical thickness is minimal (the refractive index and physical thickness of the cavity have a weak dependence on temperature). On the other hand, the lasing wavelength in edge-emitting lasers is defined by the peak-gain wavelength, which has a much stronger dependence on temperature. As a consequence, the spectral line width for high-power arrays (where heating and temperature gradients can be significant) is much narrower in VECSEL arrays than in edge-emitting-laser arrays (bar-stacks). Note that, over a 20°C change in temperature, the achieved emission wavelength in a QD VECSEL thus far can vary by less than 1.4 nm (compared to ~7 nm for edge-emitting lasers) due to the quantum effect of QDs.

- Beam quality: VECSELs operate with a circular beam, fundamental transverse TEM00 mode, and essentially diffraction-limited low beam divergence. Several factors contribute to the beam quality in VECSELs. Most important, VECSEL device external-cavity optics defines and stabilizes the circular fundamental laser transverse mode; such optical elements and their stabilization effect are not available with the more conventional edge and surface-emitting semiconductor lasers.6 If the pump spot is too small, compared to the fundamental mode size, laser threshold will be high because of the lossy un pumped regions encountered by the laser mode. If the pump spot is too large, higher order transverse laser modes with a larger transverse extent will be excited, causing multimode laser operation and thus degraded beam quality. Optimally adjusted pump spot size gives preferentially higher gain to the fundamental laser mode, while giving excess loss from the un pumped regions to the spatially wider higher order transverse modes; this stabilizes the fundamental transverse mode operation. Using pump and laser cavity optics, VECSELs have independent control allowing matching of the pump spot size and the laser fundamental transverse mode size.

Large VECSEL laser beam and pump spot sizes on the chip, tens to hundreds of microns, as compared with just a few microns for edge-emitting semiconductor lasers, contribute to the mechanical stability of the VECSEL cavity, and thus also to the stability of its fundamental transverse mode operation. Through proper cavity design VECSELs can emit in a circular single transverse mode beam (circular Gaussian). This simple beam structure greatly reduces the complexity and cost of coupling/beam-shaping optics (compared with edge-emitting lasers) and increases the coupling efficiency to the fiber or pumped medium. This has been a key advantage for the VECSEL devices in low-power laser applications. Another important factor affecting spatial beam quality and stability of VECSEL devices is the negligible thermal lensing in the thin VECSEL semiconductor chip when proper heat spreading/heat sinking is used. Thermal lensing and other beam phase profile distortions are caused by thermally induced refractive index gradients in the laser gain material. In VECSELs, thin semiconductor active region with good heat sinking implies that optical path length thermal distortions and hence beam profile changes and distortions are negligible. As a result, semiconductor VECSELs operate efficiently and with excellent beam quality across a wide range of operating power regimes from near to high above threshold.

- Higher power per unit area: VECSELs are delivering ~1350 W/cm^2 now and can deliver 2- 4 kW/cm^2 in near future, while edge-emitting lasers deliver a maximum of about 500 W/cm^2 because of gap between bar to bar which has to be maintained for coolant flow.
- Scalability, heat-sinking, and packaging: Many optically pumped VECSEL devices have been reported with power levels between 10 mw and 60 W, a range of almost four orders of magnitude, while maintaining good beam quality. Such efficient power scalability is enabled by the laser mode and pump spot-size scalability on the semiconductor VECSEL chip. Since output power of semiconductor lasers is typically limited by heat dissipation and optical intensity-induced damage, increasing beam diameter in a VECSEL helps on both accounts, distributing heat and optical power over larger beam area. For well-designed heat sinking with thin semiconductor chips, heat flow from the laser active region into heat sink is essentially one-dimensional. Therefore, increasing beam area is essentially equivalent to operating multiple lasing elements in parallel, without changing thermal or optical intensity regime of the individual lasing elements. In this scenario, both output laser power and pump power scale linearly with the active area. For high-power applications, a key advantage of VECSELs is that they can be directly processed into monolithic 2-D arrays, whereas this is not possible for edge-emitting lasers. In addition, a complex and thermally inefficient mounting scheme is required to mount edge-emitter laser bars in stacks. Good heat-sinking and packaging of VECSELs have been two key advantages. Due to the simple processing and heat-sinking technology it is much easier to package 2-D VECSEL arrays than an equivalent edge-emitting laser bar-stack. Mounting of large high-power VECSEL 2-D arrays in a "junction-down" configuration is straightforward, making the heat-removal process very efficient. In fact, VECSEL manufacturing follows the standard, well-established IC silicon industry processing.
- High temperature operation: VCSEL devices can be operated without refrigeration because they can be operated at temperatures to 80°C, The cooling system becomes very small, rugged and portable with this approach.
- Functional versatility through intracavity optical elements: Although the external optical cavity, which controls the laser transverse modes, would make these lasers more complex and requiring assembly and alignment as

compared with the simple integrated surface emitting lasers, such an external cavity of VECSELs gives tremendous versatility to VECSEL device configurations and functions. Flexible design and construction of VECSEL device cavities, such as linear two-mirror cavity, three-mirror V-shaped cavity, and four-mirror Z-shaped cavity, allow flexible insertion of intracavity optical elements. Such intracavity functional elements are very difficult to use with integrated semiconductor devices. One important option enabled by the external cavity is the insertion of intracavity spectral filters, such as Brewster's angle briefringent filters, volume gratings, or high-reflectivity gratings, to control longitudinal spectral modes of the laser and possibly to select a single longitudinal lasing mode. Moreover, the insertion of intracavity saturable absorber elements to external cavity enables VECSEL to achieve laser passive mode locking with picosecond and sub picosecond pulse generation. In this case, the length of the external short cavity allows control of the pulse repetition rates, with rates as high as 50GHz. External cavity optics also allows different beam spot sizes on the gain and absorber elements required to achieve mode locking, which controls optical intensity of the beam spots. The open cavity of VECSELs implements placement of transparent intracavity heat spreaders in direct contact with the laser gain element without thermally resistive laser mirrors in the path of heat dissipation. Since thermal management is critical for high-power VECSEL operation, the utilization of such heat spreaders tremendously broadens laser design options with chip gain, mirror, and substrate materials that do not allow effective heat removal through the on-chip mirror. Another option for VECSELs allowed by the external cavity is the microchip laser regime. Low intracavity loss of VECSELs, combined with their wide gain bandwidth, allows insertion of intracavity absorption cells, such as gas cells, for intracavity laser absorption spectroscopy (ICLAS) performing sensitive measurements of extremely weak absorption lines. Besides, availability of such high intracavity power, together with high beam quality, allows very efficient nonlinear optical operation, such as second harmonic generation, by inserting nonlinear optical crystals inside the external laser cavity. Using intracavity second harmonic generation, VECSELs have provided efficient laser output at wavelengths not accessible by other laser materials and techniques.

To summarize, these combination advantageous properties of semiconductor VECSEL devices with excellent beam quality operate efficiently across a wide range of operating power regimes from milliwatts to tens of watts with single or multiple semiconductor gain chips resulting in versatile laser applications. In their early stage application, VECSEL devices were developed as high-power single-mode fiber-coupled sources at 980nm for pumping Er-doped fiber and glass-waveguide amplifiers for optical fiber telecommunications systems. Commercial 850 nm GaAs VCSELs have been well established for these short-reach applications. Gigabit Ethernet and Fiber Channel are currently major markets for VECSELs. Long wavelength QD VECSELs emitting at 1.2–1.33 μm are currently attracting much interest for use in single-mode fiber metropolitan area and wide-area networks. Further optimization of the quantum defect and electron confinement energy is required for room temperature operation of high-performance QD VECSEL devices, the structure of which is depicted in Fig. Note that in practical application there is some trade-off between output power and beam quality in VECSELs: a multimode laser beam can better overlap the pump spot and thus produce somewhat higher output power.

Laser Diode

A laser diode is a semiconductor laser that is closely related to the light emitting diode (LED) both in form and in operation. The laser diode is quite different from the common perception of lasers as big, bulky and power-hungry devices that emit an intense beam of light that can burn or even cut. A laser diode can be considered as an LED that emits focused light and is commonly used in everyday consumer devices such as DVD and Blu-ray players, barcode scanners, fiber optic communication equipment and laser printers.

The laser diode was invented by Dr. Robert Hall of General Electric. He filed his patent on October 24, 1962, and received the patent grant on April 5, 1966, as US Patent no. 3,245,002. The diode emits laser light by bouncing photons back and forth between slices of p-type and n-type semiconductors that are roughly 1 micrometer apart. This is a similar process used in conventional lasers that produce beams by repeatedly pumping the light emitted by the atoms between two mirrors.

The forward biasing done on the semiconductors used in the laser diode forces the two charge carriers (holes and electrons) to be injected or pumped from the opposite side of the p-n junction into the other side called the depletion region, which has zero charge carriers. Because of this, laser diodes are also called injection laser diodes.

Laser Diode Construction

The laser diode is made of two doped gallium arsenide layers. One doped gallium arsenide layer will produce an n-type semiconductor whereas another doped gallium arsenide layer will produce a p-type semiconductor. In laser diodes, selenium, aluminum, and silicon are used as doping agents.

P-N Junction

When a p-type layer is joined with the n-type layer, a p-n junction is formed. The point at which the p-type and n-type layers are joined is called p-n junction. The p-n junction separates the p-type and n-type semiconductors.

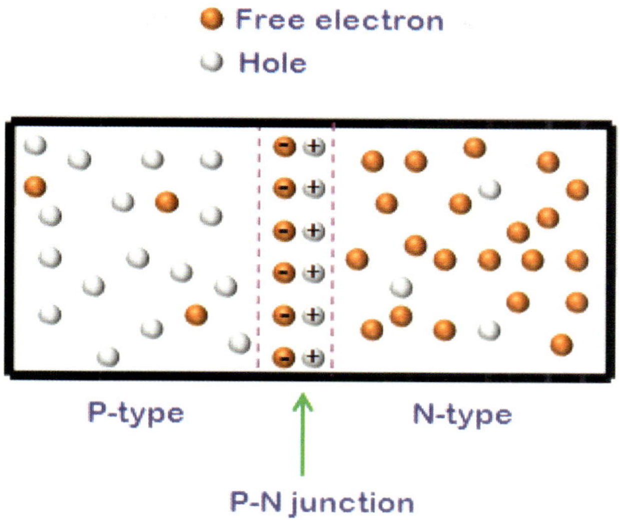

For the construction of laser diodes, gallium arsenide is chosen over silicon. In silicon diodes, the energy is released during recombination. However, this release of energy is not in the form of light.

In gallium arsenide diodes, the release of energy is in the form of light or photons. Therefore, gallium arsenide is used for the construction of laser diodes.

N-type Semiconductor

Adding a small percentage of foreign atoms into the intrinsic semiconductor produces an n-type or p-type semiconductor.

If pentavalent impurities are added to the intrinsic or pure semiconductor, an n-type semiconductor is produced. In n-type semiconductors, free electrons are the majority charge carriers whereas holes are the minority charge carriers. Therefore, free electrons carry most of the electric current in n-type semiconductors.

P-type Semiconductor

If trivalent impurities are added to the pure semiconductor, a p-type semiconductor is produced. In p-type semiconductors, holes are the majority charge carriers whereas free electrons are the minority charge carriers. Therefore, holes carry most of the electric current in p-type semiconductors.

Main Steps Required for Producing a Coherent Beam of Light in Laser Diodes

The main steps required for producing a coherent beam of light in lasers diodes are: light absorption, spontaneous emission, and stimulated emission.

Absorption of Energy

Absorption of energy is the process of absorbing energy from the external energy sources.

In laser diodes, electrical energy or DC voltage is used as the external energy source. When the DC voltage or electrical energy supplies enough energy to the valence electrons or valence band electrons, they break bonding with the parent atom and jumps into the higher energy level (conduction band). The electrons in the conduction band are known as free electrons.

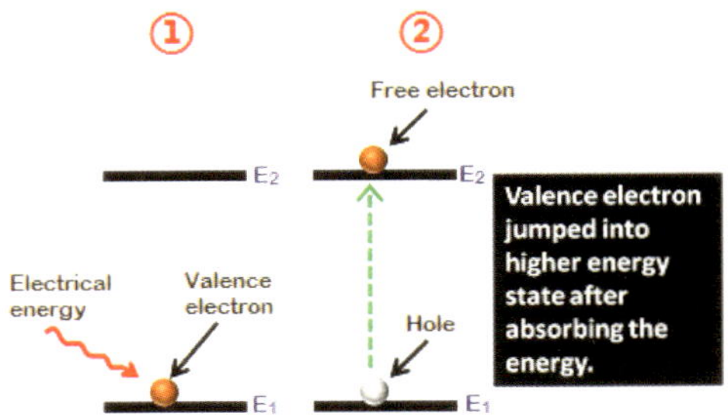

Absorption

When the valence electron leaves the valence shell, an empty space is created at the point from which electron left. This empty space in the valence shell is called a hole.

Thus, both free electrons and holes are generated as a pair because of the absorption of energy from the external DC source.

Spontaneous Emission

Spontaneous emission is the process of emitting light or photons naturally while electrons falling to the lower energy state.

In laser diodes, the valence band electrons or valence electrons are in the lower energy state. Therefore, the holes generated after the valence electrons left are also in the lower energy state.

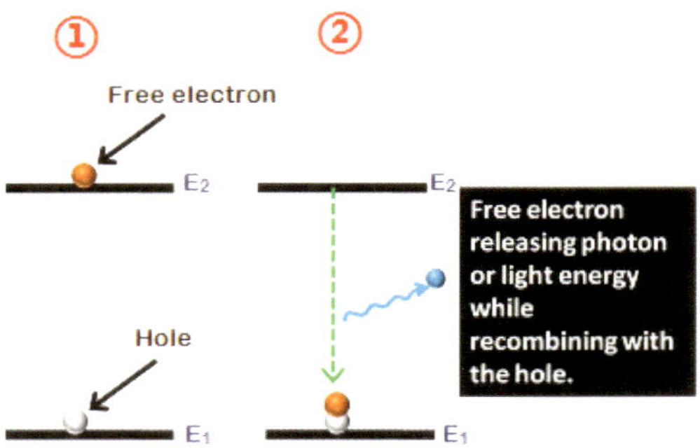

Spontaneous emission

On the other hand, the conduction band electrons or free electrons are in the higher energy state. In simple words, free electrons have more energy than holes.

The free electrons in the conduction band need to lose their extra energy in order to recombine with the holes in the valence band.

The free electrons in the conduction band will not stay for long period. After a short period, the free electrons recombine with the lower energy holes by releasing energy in the form of photons.

Stimulated Emission

Stimulated emission is the process by which excited electrons or free electrons are stimulated to fall into the lower energy state by releasing energy in the form of light. The stimulated emission is an artificial process.

In stimulated emission, the excited electrons or free electrons need not wait for the completion of their lifetime. Before the completion of their lifetime, the incident or external photons will force the free electrons to recombine with the holes. In stimulated emission, each incident photon will generate two photons.

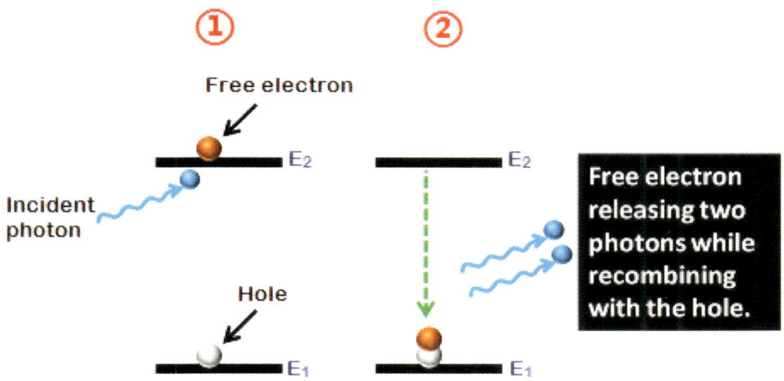

Stimulated emission

All the photons generated due to the stimulated emission will travel in the same direction. As a result, a narrow beam of high-intensity laser light is produced.

Working of Laser Diode

When DC voltage is applied across the laser diode, the free electrons move across the junction region from the n-type material to the p-type material. In this process, some electrons will directly interact with the valence electrons and excites them to the higher energy level whereas some other electrons will recombine with the holes in the p-type semiconductor and releases energy in the form of light. This process of emission is called spontaneous emission.

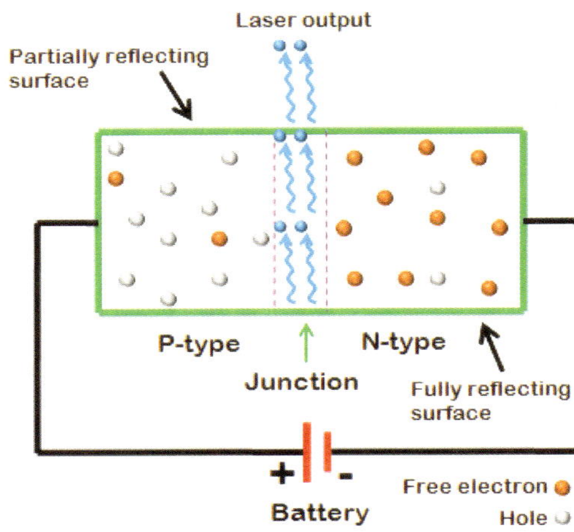

Laser diode

The photons generated due to spontaneous emission will travel through the junction region and stimulate the excited electrons (free electrons). As a result, more photons are released. This process of light or photons emission is called stimulated emission. The light generated due to stimulated emission will moves parallel to the junction.

The two ends of the laser diode structure are optically reflective. One end is fully reflective whereas another end is partially reflective. The fully reflective end will reflect the light completely whereas the partially reflective end will reflect most part of the light but allows a small amount of light.

The light generated in the p-n junction will bounce back and forth (hundreds of times) between the two reflective surfaces. As a result, an enormous optical gain is achieved.

The light generated due to the stimulated emission is escaped through the partially reflective end of the laser diode to produce a narrow beam laser light.

All the photons generated due to the stimulated emission will travel in the same direction. Therefore, this light will travel to long distances without spreading in the space.

Advantages of Laser Diodes

1. Simple construction
2. Lightweight
3. Very cheap
4. Small size
5. Highly reliable compared to other types of lasers
6. Longer operating life

7. High efficiency
8. Mirrors are not required in the semiconductor lasers
9. Low power consumption.

Disadvantages of Laser Diodes

1. Not suitable for the applications where high powers are required.
2. Semiconductor lasers are highly dependent on temperature.

Applications of Laser Diodes

1. Laser diodes are used in laser pointers.
2. Laser diodes are used in fiber optic communications.
3. Laser diodes are used in barcode readers.
4. Laser diodes are used in laser printing.
5. Laser diodes are used in laser scanning.
6. Laser diodes are used in range finders.
7. Laser diodes are used in laser absorption spectrometry.

Quantum Well Laser

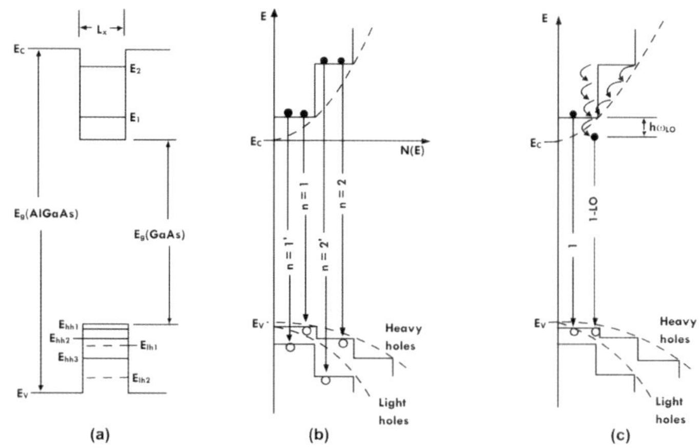

Figure: (a) Quantum well laser structure showing discrete energy levels.
(b) Excitation of electrons happen to these discrete levels (c) Recombination leads to emission of light. Some of the re combinations are also mediated by the lattice vibrations

The active layer in a hetero junction laser can be made very thin. If the width of the active layer is comparable to the de Broglie wavelength of the electrons (typically around the nm range) then quantum confinement can be achieved. The energy levels are no longer continuous but are quantized into discrete values. This is shown in figure above. The spacing between the quantized energy levels depend on the width of the active layer. This also controls the wavelength of the laser radiation. If the confinement is along only one direction (thickness) then it is a quantum well laser.

Quantum well lasers have active regions of about 100 Å thick, which restricts the motion of the carriers (electrons and holes) in a direction normal to the well. This results in a set of discrete energy levels and the density of states is modified to a two-dimensional-like density of states. This modification of the density states results in several improvements in lasers characteristics such as lower threshold current, higher efficiency, and higher modulation bandwidth and lower CW and dynamic spectral width. All of these improvements were first predicted theoretically and then demonstrated experimentally.

Quantum well lasers require fewer electrons and holes to reach threshold than conventional double hetero structure lasers. A well-designed quantum well laser can have an exceedingly low threshold current.

Moreover, since quantum efficiency (photons-out per electrons-in) is largely limited by optical absorption by the electrons and holes, very high quantum efficiencies can be achieved with the quantum well laser.

To compensate for the reduction in active layer thickness, a small number of identical quantum wells are often used. This is called a multi-quantum well laser.

The development of In GaAsP quantum well lasers was made possible by the development of MOCVD and GSMBE growth techniques. The transmission electron micrograph (TEM) of a multiple Quantum Well laser structure is shown in the following figure. It shows five In Ga As quantum wells grown over n-InP substrate. The well thickness is 70 Å, and the wells are separated by barrier layers of InGaAsP λ = 1.1 um. Multi quantum well lasers with threshold current densities of 600 A/cm^2 have been fabricated.

The laser has a Multi-Quantum-Well (MQW) active region and it utilizes Fe doped InP semi-insulating layers for current confinement and optical confinement.

The threshold current at room temperature is ~ 1.1 mA. The laser is 170 um long and has 90% and 70% reflective coating at the facets. Such low-threshold lasers are important for array applications.

Recently, QW lasers were fabricated that have higher modulation bandwidth than regular DH lasers. The current confinement and optical confinement in this laser are carried out using MOCVD grown Fe doped InP lasers. The laser structure is then further modified by using a small contact pad and etching channels around the active region mesa. These modifications are designed to reduce the capacitance of the laser structure. A 3-dB bandwidth of 25 GHz is obtained.

Transparency of Current Density

The transparency current density represents a fundamental limit to achieving the lowest lasing threshold for semiconductor lasers in general.

The current needed for lasing is composed of two parts: the first part being the current needed for maintaining the electron density at the optical transparency level, and beyond that a second part to attain the necessary gain to overcome all the losses in the laser cavity. It can be argued (and can actually be demonstrated experimentally) that a laser cavity can be designed such that the losses are minimal, but this can only reduce the second part of the threshold current while the first part, that responsible for optical transparency, is unaffected.

The key to building an ultralow threshold laser is thus to design a laser cavity with a very low loss, with a material that has the lowest transparency current density. A single quantum well structure is one that possesses both of these qualities and, when combined with high reflectivity coatings to minimize mirror loss, results in some of the lowest lasing threshold currents achieved to date.

Strained Quantum-well Lasers

Quantum well lasers have also been fabricated using an active layer whose lattice constant differs slightly from that of the substrate and cladding layers. Such lasers are known as strained quantum-well lasers.

Over the last few years, strained quantum well lasers have been extensively investigated all over the world. They show many desirable properties such as:

1. A very low-threshold current density;
2. A lower line width than regular Multi-Quantum-Well (MQW) lasers both under continuous wave (CW) operation and under modulation.

The origin of the improved device performance lies in the band-structure changes induced by the mismatch-induced strain. Strain splits the heavy-hole and the light-hole valence bands at the T point of the Brillouin zone where the bands gap is minimum in direct band-gap semiconductors.

Two material systems have been widely used for strained quantum well lasers:

1. In GaAs grown over InP by the MOCVD or the CBE growth technique;
2. In GaAs grown over GaAs by the MOCVD or the MBE growth technique.

The first material system is of importance for low-chirp semiconductor laser for light wave system applications. The second material system has been used to fabricate high-power lasers emitting near 0.98um, a wavelength of interest for pumping erbium-doped fiber amplifiers (EDFA).

Inter Band Cascade Laser

Inter band cascade lasers (ICLs) are a type of laser diode.

The inter band cascade laser (ICL) is a promising mid-IR (defined here as l = 3-6 μm) source that combines the relatively long upper-level recombination lifetime of a conventional diode laser with the voltage-efficient cascading scheme introduced by the quantum cascade laser (QCL, which produces coherent light via inter sub band transitions within the conduction band). The mid-IR spectral region has gained increasing relevance in recent years, because it hosts numerous "fingerprint" spectral lines that can be used to sense trace gases such as methane, carbon dioxide, carbon monoxide, formaldehyde, etc. While practical sensing systems typically require cw emission into a single spectral mode, a low output power on the order of 1 mW is usually sufficient. On the other hand, such military applications as infrared countermeasures require substantial cw output powers, whereas spectral purity is generally less critical. Other potential applications for compact and efficient mid-IR sources include industrial process control, clinical breath analysis, free-space optical communications, IR scene projection, and the detection of chemical/biological threats.

Nonetheless, until 2002 no semiconductor source could offer the continuous-wave (cw) operation at ambient temperature that is essential if any mid-IR technology is to be considered sufficiently practical for widespread implementation. Earlier development efforts had focused on extending the we'll established diode laser architecture, with type-I alignment of the conduction and valence bands, beyond the 2-3 μm spectral window by increasing the strain in GaSb-based multiple quantum wells (MQWs). However, significant challenges included rapid wavelength scalings of the Auger non-radiative decay and free-carrier absorption loss, carrier escape associated with a marginal valence-band offset in the MQWs, and the general immaturity of GaSb-based growth and processing technologies.

This view was upended by the rapid progress of QCLs grown on InP. A modest temperature sensitivity, coupled with efficient heat dissipation, have enabled QCL output powers to reach 5 W cw in a near-diffraction-limited beam at room temperature (RT). The primary disadvantages are the QCL's high threshold current density, associated with rapid phonon-assisted depopulation of the upper lasing sub band and interface roughness scattering ($\approx$ 1 ps), coupled with a bias of at least $\approx$ 10 V because 30-50 stages are needed to supply sufficient gain.

Even though the ICL was invented in 1994, the same year that the QCL was first demonstrated experimentally, over the intervening period it has received much less attention and far fewer resources have been devoted to its development. Within 2-3 years, the hole injector was added to the concept, along with replacement of the single type-II interface in the active gain region with a "W" configuration that incorporates two InAs electron QWs on both sides of the GaInSb hole QW. The "W" structure has become a standard feature of state-of-the-art ICL designs, even though its advantage is not so clear if the non radiative recombination rate decreases with electron-hole wave function overlap even as the gain increases.

While the first experimental realization of an ICL was reported in 1997, the initial devices operated only at cryogenic temperatures due to several design aspects that were far from what is now known to be optimal. In particular, excessive electron injector thicknesses induced high threshold current densities and low external differential quantum efficiencies (EDQEs) per stage. Nevertheless, gradual improvements in both designs and the growth techniques led to steady improvement of the temperature performance, with Tmax = 250-286 K being reached for pulsed operation in the 1998-2000 time frame. A significant milestone was the demonstration in 2002 of pulsed operation at RT by an ICL with 18 active stages, even though the threshold current density (Jth) exceeded 6 kA/cm^2. Designs employing the "W" active QW configuration subsequently operated at = 3.3 μm with much lower RT Jth (≈ 1 kA/cm^2), along with cw lasing to T = 200 K.

In 2005, the group at Jet Propulsion Laboratory (JPL) substantially reduced the pulsed RT Jth further, to 630 A/cm^2, with only 12 active stages. This eventually led to cw operation up to Tmax = 264 K, which enabled JPL to qualify single-mode distributed-feedback (DFB) ICLs for methane detection on the NASA Mars Curiosity Mission.

The first ICLs designed, grown, processed, and characterized at Naval Research Laboratory (NRL) were reported in 2006. In subsequent NRL designs the number of stages was reduced from 10 to 5, in order to minimize the threshold power density. Simulations had determined that an ICL's gain per unit current density is actually quite high, e.g., much higher than in a QCL that typically requires 30 or more stages to minimize Jth at RT. NRL subsequently reported a series of advances in such key performance figures of merit as pulsed Jth, Pth, and EDQE, at and above RT, by employing a much thinner electron injector and two GaSb QWs in the hole injector. The result was a lower RT threshold current density of 1.15 kA/cm^2 for a 10-stage structure. An epitaxial-side-up narrow ridge from that wafer, with Au electroplating deposited on top for heat dissipation, operated cw to Tmax = 269 K.

In 2008, an NRL ICL emitting at = 3.75 μm became the first to operate cw at RT. That milestone was made possible by several additional design modifications, including the introduction of lightly n-doped GaSb separate-confinement layers (SCLs) to reduce the waveguide loss, decrease of the n doping in the InAs/AlSb optical cladding layers, and reduction of the number of stages to five. The resulting pulsed Jth at RT dropped to ≈ 400 A/cm^2, the lowest threshold reported up to that point.

In the past 6 years, NRL has continued to explore numerous other design modifications. While some had little or no significant impact on the laser performance, others were quite beneficial to the threshold and efficiency. This applies especially to the "carrier rebalancing" concept. In parallel, the University of Würzburg (UWUERZ) and Nano plus GmbH have reported a variety of valuable recent contributions to the ICL development. While alternative approaches that are suitable for the longest wavelengths of the mid-IR and beyond are now being explored by Prof. Rui Yang and colleagues at the University of Oklahoma, RT cw operation has not yet been reported.

Distributed Bragg Reflector Laser

A distributed Bragg reflector laser is a laser, where the laser resonator is made with at least one distributed Bragg reflector (DBR) outside the gain medium (the active region). A DBR is a Bragg mirror, i.e., a light-reflecting device (a mirror) based on Bragg reflection at a periodic structure. In most cases, the Bragg mirror is more specifically a quarter-wave mirror, providing the maximum amount of reflection for the given number of layers.

Lasers of DBR type are usually laser diodes, but the term is also sometimes used for fiber lasers containing fiber Bragg gratings. Both laser types are described below. Most solid-state bulk lasers actually also use laser mirrors which are Bragg mirrors; nevertheless such lasers are not called DBR lasers.

A DBR laser is different from a distributed feedback laser, where the whole active medium is embedded in a single distributed reflector structure.

DBR Laser Diodes

A DBR laser diode contains some corrugated waveguide structure (a grating section) providing wavelength-dependent feedback to define the emission wavelength. Another section of the laser waveguide acts as the amplifying medium (active region), and the other end of the resonator may have another DBR.

DBR laser diodes are usually single-frequency lasers with diffraction-limited output, and often they are wavelength-tunable (→ tunable lasers). Tuning within the free spectral range of the laser resonator may be accomplished with a separate phase section, which can e.g. be electrically heated, or simply by varying the temperature of the gain region via the drive current. If the temperature of the whole device is varied, the wavelength response is significantly smaller than for an ordinary single-mode laser diode, since the reflection band of the grating is shifted less than the gain maximum. Electro-optic tuning can also be accomplished. Mode-hop free tuning over a larger wavelength region is possible by coordinated tuning of the Bragg grating and the gain structure.

There are more sophisticated device designs, exploiting a kind of Vernier effect with sampled gratings (SG-DBR laser), that offer a tuning range as wide as e.g. 40 nm, although not without mode hops.

The line width of a DBR diode is typically a few megahertz. Due to the relatively short laser resonator, it is larger than that e.g. of an external-cavity diode laser.

There are MOPA structures where an additional amplifier section (a semiconductor optical amplifier) is placed on the same semiconductor chip. The actual DBR laser is then the seed laser. Output powers well above 100 mW can be achieved with such devices. It is also possible to directly generate high powers (> 10 W) with a broad-area laser diode having a surface Bragg grating within its resonator. The emission may then no longer be in a single mode, but still with a relatively small bandwidth.

Vertical cavity surface-emitting lasers (VCSELs) are actually also distributed Bragg reflector lasers, even though the term "DBR laser diodes" is normally used for edge-emitting semiconductor lasers.

Applications of DBR laser diodes include optical fiber communications, free-space optical communications, laser cooling, optical metrology and sensors, and high-resolution spectroscopy. DBR lasers actually compete with external-cavity diode lasers (ECDLs), which also offer wavelength-tunable single-frequency output, with potentially better performance e.g. in terms of noise, but also requiring a significantly more complex setup. Chips containing DBR laser arrays can serve as very compact sources for use in wavelength division multiplexing systems.

DBR Fiber Lasers

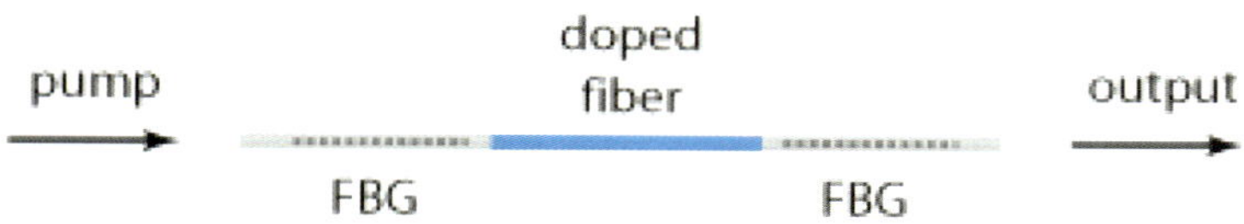

Figure: Short DBR fiber laser for narrow-line width emission

A fiber laser of DBR type usually has a linear laser resonator formed by an active (rare-earth-doped) fiber between two fiber Bragg gratings. Compared with a fiber DFB laser, which consists of a single grating in a fiber with laser gain, a DBR fiber laser has a longer laser resonator and thus the potential for higher output power, higher power efficiency, and narrower line width. On the other hand, this can also lead to less robust single-frequency operation, or to multimode operation with a correspondingly much larger emission bandwidth. Single-frequency DBR fiber lasers offer similar output powers as DBR laser diodes: tens of milliwatts or sometimes > 100 mW.

Sampled grating designs, can also be used in DBR fiber lasers. The tuning range achievable can again have a width of tens of nanometers.

Distributed Feedback Laser

A distributed feedback laser (DFB) is a type of laser diode, in which the feedback is produced by light waves reflection from a periodic lattice generated in the active medium.

In the injection hetero lasers with distributed feedback (DFB lasers) to create a feedback one of the hetero interfaces is corrugated, which creates a periodic variation in the refractive index and leads to an interference reflection. The lattice period is selected so as to ensure satisfaction of the Bragg condition for the reflections in the opposite direction:

$$2n_e d = \lambda,$$

Where, λ is the wavelength of light in vacuum, n_e is the effective refractive index of the active region, d is the lattice period. The lattice period is of the order of 100 nm (in gallium arsenide laser, for example, it is 130 nm).

The condition of reflection from the periodic structure is satisfied for the beams propagating in both directions. Thus, the periodic lattice creates feedback in both directions, distributed along the entire length of the laser. Since the feedback produced by the periodic lattice is selective, the DFB laser operates in the single-mode generation regime.

DFB lasers can be placed directly onto the surface of a semiconductor substrate and connected to waveguides on the surface of the substrate to create photonic integrated circuits.

The most effective alignment between a DFB laser and a channel waveguide is achieved in DFB lasers with separate optical and electron confinement in double hetero structures (SCDH DFB lasers).

DFB lasers are characterised by temperature stability of the oscillation frequency, which is uniquely determined by the optical lattice period. The temperature dependence coefficient of the emission wavelength in a typical DFB laser is 0.1 nm/deg; it is determined by the temperature dependence of the refractive index. This makes it possible to tune the radiation frequency by adding a temperature control unit to the laser. Such simplicity of design is a very significant advantage of tunable DFB lasers. Yet, the major drawback of these lasers is the limited frequency tuning range.

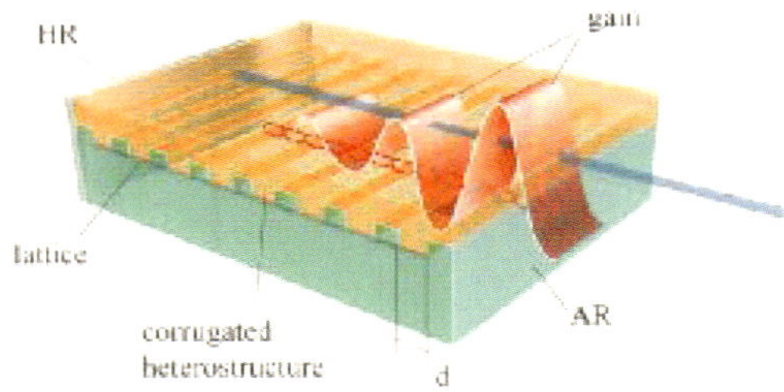

Optical arrangement of a distributed feedback semiconductor laser.
HR — high reflection ratio mirror; AR — antireflection coating

Integrated one-piece designs containing a grid of multiple DFB-lasers combined in one unit allow the tuning range to be increased. Thus, a block of eight parallel DFB lasers integrated by a multi-mode waveguide coupler (MMI) enables a tuning range of up to 60 nm. To increase the power output of the tunable laser it is equipped with an amplifier. Various designs of multi-element tunable lasers with the number of elementary DFB lasers in the lattice ranging from 2 to 8 have already been developed and studied.

References

- Semiconductor-lasers: rp-photonics.com, Retrieved 11 May 2018
- About-semiconductor-laser-diode: fiberlabs-inc.com, Retrieved 19 March 2018
- Semiconductor-laser, microwave-radar, electronics: daenotes.com, Retrieved 25 April 2018
- Quantum-cascade-laser-basics: teamwavelength.com, Retrieved 15 July 2018
- Laser-diode-14955: techopedia.com, Retrieved 31 May 2018
- What-is-quantum-well-laser-95044678: fiberoptics4sale.com, Retrieved 30 March 2018
- Distributed-bragg-reflector-lasers: rp-photonics.com, Retrieved 31 April 2018

6

Laser Devices and Applications

There has been significant development in the field of laser technology. The aim of this chapter is to explore the diverse laser devices, such as laser tracker, 3D scanner, laser projector and pointer and their applications in printing, scanning, use in medicine, etc.

Laser Tracker

Laser trackers provide fast measurement of a target which can be moved almost anywhere within line-of-sight of the base unit. A key factor in favour of the tracker is the high relative accuracy that can be achieved.

Some laser trackers can measure object features up close and as far away as 60 meters. Some have typical single point accuracy of about 0.001" (0.025 mm) at distances to several meters.

As a consequence laser trackers have largely supercoded more traditional methods such as theodolites or conventional metrology tools such as collimators.

Uses of Laser Trackers

- Robot tracking, calibration, maintenance and testing;
- Shipbuilding;
- Aircraft manufacturing;
- Automotive jig build and set up;
- Verification of the design of manufactured structures;
- Reverse engineering;
- Inspection and alignment.

Type of Information Gathered

- Raw 3-D co-ordinates;
- CAD models;

- Surfaces;
- Deformation;
- Movement;
- Reverse engineering data;
- Tracking of objects.

Laser Tracker Working

The operation of a laser tracker is easy to understand: It measures two angles and a distance. The tracker sends a laser beam to a retro-reflective target held against the object to be measured. Light reflected off the target retraces its path, re-entering the tracker at the same position it left. Retro-reflective targets vary, but the most popular is the spherically mounted retro-reflector (SMR). As light re-enters the tracker, some of it goes to a distance meter that measures the distance from the tracker to the SMR. The distance meter may be either of two types, interferometer or absolute distance meter (ADM).

A laser tracker contains two angular encoders. These devices measure the angular orientation of the tracker's two mechanical axes: the azimuth axis and the elevation (or zenith) axis. The angles from the encoders and the distance from the distance meter are sufficient to precisely calculate the center of the SMR. Because of the spherical design of the SMR, its center is always at a fixed offset distance with respect to any surface being measured, so the coordinates of surfaces or points measured with the SMR are easily obtained. Distance measurement, an important function of the laser tracker, can be either incremental or absolute. Incremental distance measurement is made with an interferometer and a frequency-stabilised, helium-neon laser.

The laser light splits into two beams. One travels directly into the interferometer. The other beam travels out of the tracker, reflects off the SMR and, on the return path, passes into the interferometer. Inside the interferometer, the two beams of light interfere, resulting in a cyclic change each time the SMR moves closer to or farther from the tracker by a distance equal to one quarter of the light's wavelength (~0.0158 micron). Electronic circuitry counts the cyclic changes (known as "fringe counts") to determine the distance traveled.

In a typical measurement sequence, the operator places the SMR in the tracker's home position and resets the interferometer to the known (home) distance. As the operator moves the SMR to the desired location, the laser tracks along, remaining fixed to the centre of the SMR. This procedure works well as long as the beam from the tracker to the SMR isn't broken by an obstruction in the beam path. If the beam is broken, however, the number of counts is no longer valid and the distance isn't known. When this happens, the tracker signals that an error has occurred. The operator must then return the SMR to a reference point, such as the tracker's home position.

Absolute distance measurement capability has been around for a long time. Within the last 10 years, however, ADM systems have undergone dramatic improvement, offering accuracy comparable to interferometers.

The advantage of ADM measurement over incremental distance measurement is the ability simply to point the beam at the target and shoot. The ADM system measures the distance to the target automatically, even if the beam has previously been broken. In a tracker with ADM, infrared light from a semiconductor laser reflects off the SMR and re-enters the tracker, where it's converted into an electrical signal. Electronic circuitry analyses the signal to determine its time of flight, multiplying this value by the speed of light in air to determine the distance from the tracker to the SMR.

Absolute-distance meters first appeared in laser trackers in the mid-1990s. At that time, ADM units measured too slowly to permit scanning of surfaces. Because of this, all early laser trackers contained either an interferometer alone or an interferometer and an ADM. Today, some absolute-distance meters have been made fast enough to permit high-speed scanning with negligible loss in accuracy. Hence, some modern trackers contain only an ADM with no interferometer.

Another tracker function is beam steering and control. One type of tracker launches the laser beam directly from its rotating structure. Another type of tracker reflects a laser beam off a rotating mirror. In either case, the tracker points the laser beam in the desired direction by rotating the mechanical axes. In many applications, the tracker keeps the beam centered on a rapidly moving SMR. It accomplishes this by splitting off part of the returning laser beam to a position-sensing detector (PSD). If the laser beam strikes the SMR off center, the split-off beam also strikes the PSD off center, creating an error signal. This signal controls the mechanical axes' rotation to keep the beam centred on the SMR.

Measuring Coordinates with the Tracker

Trackers collect 3-D coordinate data, which can be "fit" by software to geometrical entities such as points, planes, spheres, or cylinders. Usually, the data are displayed within a local coordinate system tied to an object's features. A flat surface on the object, for example, may represent the x-y plane. Alternatively, the local coordinate system may be established by features that represent points or lines. Points may be represented by tooling holes, into which target nests or tooling balls are inserted.

Benefits

- Intuitive: operator places the target anywhere a co-ordinate is required;
- Fast: each data point can be recorded in a few seconds;
- Single user: one device and operator can record points working alone;

- Range: typically tens of metres, creating a large working volume;
- Reasonably: large installed user base in high value operations.

Limitations

- Occlusion: can only operate in line of sight; breaking the beam requires resetting the co-ordinate system;
- Contact: target must physically touch the measured point;
- Offset: recorded co-ordinates are offset from the actual surface;
- Target size: size of the retro-reflector limits the minimum radius of curvature measurable;
- Static scene: scene must remain static as the points measured;
- Environment: changes in air temperature, pressure and humidity can affect measurements;
- Cost: a Laser Tracker is an expensive piece of equipment;
- Portability: the Tracker is relatively large and heavy making it unsuitable for some applications;
- Ruggedness: the Tracker is a high precision piece of equipment and is unsuitable for use in many hazardous, dirty or unstable environments.

3D Scanner

A 3-D scanner is an imaging device that collects distance point measurements from a real-world object and translates them into a virtual 3-D object.

3-D scanners are used for creating life-like images and animation in movies and video games. Other applications of 3-D scanning include reverse engineering, prototyping, architectural and industrial modeling, medical imaging and medical device modeling. 3-D printers can use data from 3-D scans to create physical objects.

Contact-based 3-D scanners work in a number of ways. One type has a carriage system and a control flatbed on which the object rests, while arms take measurements. Another type has articulated arms, and measurements are calculated from the joint angles of those arms.

Optical 3-D scanners use photographic, stereoscopic cameras, lasers or structured or modulated light. Optical scanning often requires many angles or sweeps.

Laser-based methods use a low-power, eye-safe pulsing laser working in conjunction with a camera. The laser illuminates a target, and associated software calculates the time it takes for the laser to reflect back from the target to yield a 3-D image of the scanned item.

Non-laser light-based scanners use either light that is structured into a pattern or a constantly modulated light and then record the formation the scanned object makes. Some medical tomographic scanners use X-rays to create a 3-D X-ray scan image.

Commercial desktop and hand held 3-D scanners vendors include digitizer, Next engine, go SCAN and Fuel 3D. Most products use a combination of the techniques to make a faster and more accurate device that can capture color as well as form.

3D Scanning Process

An object that is to be laser scanned is placed on the bed of the digitizer. Specialized software drives the laser probe above the surface of the object. The laser probe projects a line of laser light onto the surface while 2 sensor cameras continuously record the changing distance and shape of the laser line in three dimensions (XYZ) as it sweeps along the object.

Resulting Data

The shape of the object appears as millions of points called a "point cloud" on the computer monitor as the laser moves around capturing the entire surface shape of the object. The process is very fast, gathering up to 750,000 points per second and very precise (to ±.0005″).

Modeling Choice Depends on Application

After the huge point cloud data files are created, they are registered and merged into one three-dimensional representation of the object and post-processed with various software packages suitable for a specific application.

Point Cloud Data for Inspection

If the data is to be used for inspection, the scanned object can be compared to the designer's CAD nominal data. The result of this comparison process is delivered in the form of a "color map deviation report," in PDF format, which pictorially describes the differences between the scan data and the CAD data.

CAD Model for Reverse Engineering

Laser scanning is the fastest, most accurate, and automated way to acquire 3D digital data for reverse engineering. Again, using specialized software, the point cloud data is used to create a 3D CAD model of the part's geometry. The CAD model enables the precise reproduction of the scanned object, or the object can be modified in the CAD model to correct imperfections. Laser Design can provide a surface model or the more complex solid model, whichever results are needed for the application.

3D Scanning Technologies

Laser Triangulation

Laser triangulation based 3D scanners use either a laser line or a single laser point to scan across an object.

The laser is first casted by the 3D scanner. As the laser light reflects off the scanned object, its initial trajectory is modified and picked up by a sensor.

From the modification of the laser trajectory and trigonometric triangulation, the system can discern a specific deviation angle. The calculated angle is directly linked to the distance from the object to the scanner. When the 3D scanner collects enough distances, it is capable of mapping the surface's object and to create a 3D scan.

The main advantages of the laser triangulation technology for 3D scanning are its resolution and accuracy.

One downside of the laser triangulation technology is its sensibility to the properties of the surface to 3D scan. Very shiny or transparent surfaces are particularly problematic.

Structured Light

Structured light 3D scanners use trigonometric triangulation but do not rely on a laser. Instead, the structured light 3D scanning technology works with the projection of a series of linear patterns onto an object.

The system is then capable to examine the edges of each line in the pattern and to calculate the distance from the scanner to the object's surface.

The structured light used for 3D scanning can be white or blue and generated by numerous types of projectors, such as Digital Light Processing (DLP) technology. The projected pattern is usually a series of light rays but can also be a randomized dot matrix.

Compare Structured Light 3D Scanners

The main advantages of the structured light technology for 3D scanning are its speed, resolution and ability to 3D scan people.

One downside of the structured light technology is its sensibility to lighting conditions and issues to work outside.

Photography – Photogrammetry

Photogrammetry is the science of making measurements from photographs, especially for recovering the exact positions of surface points. Photogrammetry is based on a mix of computer vision and powerful computational geometry algorithms.

The principle of photogrammetry is to analyze several photographs of a static subject, taken from different viewpoints, and to automatically detect pixels corresponding to a same physical point.

The data input required from the user are the parameters of the camera such as focal length and lens distortion.

The main challenge for this 3D scanning technology is to analyse many photos and thousands of points with high accuracy. A very powerful computer is required to run photogrammetry algorithms.

The main advantages of the photogrammetry technology for 3D scanning are its precision and acquisition speed. The photogrammetric technology is also capable to reconstruct subjects of various scales, photographed from the ground or from the air.

One downside of the photogrammetry technology is its sensibility to the resolution of the input photographs and the time it takes to run the algorithms.

Contact Based 3D scanning

Contact based 3D scanning is also known as digitizing. The contact technology for 3D scanning implies a contact based form of 3D data collection.

Contact 3D scanners probe the subject through physical touch, while the object is firmly hold in place. A touching probe is moved on the surface to various points of the object to record 3D information.

The probe is sometimes attached to a articulated arm capable of collecting all its respective configurations and angles for more precision. Some specific configurations of contact based 3D scanners are called Coordinated Measuring Machines (CMM).

Compare Contact 3D Scanners

Contact 3D scanning is widely used for performing quality control of parts after fabrication or during maintenance operations.

The main advantages of the contact technology for 3D scanning are its precision and ability to scan transparent or reflective surfaces.

The downsides of the contact 3D scanning technology is its speed and inadequacy to work with organic freeform shapes.

Laser Pulse-based 3D scanners

The Laser pulse-based 3D scanners, also known as time-of-flight scanners or Lidar, measure how long a casted laser takes to hit an object and come back.

Because the speed of light is exactly known, the time it takes for the laser to do the way back trip gives the exact distance between the 3D scanner and the object.

In order to measure precisely the distance, the 3D scanner requires to compute millions of laser's pulses with a picosecond (0.000000001 second!) accuracy.

Because each measure only collects one point, the 3D scanner needs to cast its laser 360 degrees around it. To perform this feature, the 3D scanner is usually fitted with a mirror which changes the orientation of the laser.

Time of flight 3D scanners encompass both laser pulse and phase shift lasers. Phase shift laser 3D scanners are a sub-category of laser pulse 3D scanners. In addition to pulsing the laser, the phase shift systems also modulate the power of the laser beam. The phase shift lasers offer a better overall performance.

The main advantage of the laser pulse 3D scanners is their ability to 3D scan very big objects and environments. They are however quite slow.

Applications

Reverse Engineering

Catering to a wide variety of 3D scanning applications in multiple industries, Artec 3D constantly develops and improves 3D scanning solutions that are perfect for use in fast prototyping and reverse engineering. Our highly precise handheld 3D scanners are a fantastic solution that allows both new and experienced users to effortlessly capture the required data in no time at all, whatever the application. Portable, lightweight and easy to use, our professional handheld 3D scanners are guaranteed to improve your workflow and provide accurate results every time, saving you both time and money.

Industrial Design and Manufacturing

Developing new product designs, taking the measurements of objects with complex geometry or automating workflow at manufacturing facilities used to take days or weeks to complete. Now 3D scanning technology can be used for these kinds of applications. You can use Artec 3D scanners to quickly capture just about any industrial object, from a small mechanical part to a turbine, with astonishing accuracy. The resulting 3D model can then be exported to a variety of CAD and CAM programs and from there gauged and modified to improve the product's design and performance or integrate it into a new production system.

Healthcare

Designing customized prosthetic and orthotic devices requires real precision and paying close attention to the patient's individual anatomy. This is one of the many healthcare applications, for which Artec 3D scanners prove to be indispensable tools. They are successfully used to produce quick and accurate scans of the body or body parts, even if the person moves about a little during scanning. Artec scanners use structured light and pose no risk to the health of the person scanned. They easily capture the geometry of the body and don't need any reference targets to be attached to the patient's body.

Science and Education

Universities, colleges and laboratories are embracing 3D scanning as a powerful tool that allows students and researchers to study artifacts in greater detail than ever before without risking damaging them. The world's leading museums also use Artec 3D scanners to digitize artifacts and create online galleries, facilitating access to their collections for art specialists and academics no matter where they are based. Unlike the original objects that can be seen only from behind the glass at museums where they are exhibited, their 3D copies can be easily viewed from any angle and in different resolutions.

Art and Design

3D scanning opens up limitless possibilities for artists of any kind, enabling them to translate their most fantastic ideas into reality. The technology drives forward the movie industry and video games – many stunts and visual effects would be difficult or even impossible to bring off before the advent of 3D scanning. Artec has perfected the technology to produce 3D scanners that have become the preferred choice for making top-grossing Hollywood blockbusters, such as World War Z, Terminator Genisys, Jurassic World and many more. In addition to helping create new pieces of art, Artec also takes care of ancient artifacts and monuments by digitizing them in 3D to preserve their exact copies for posterity as the objects themselves may disintegrate over time.

Laser Projector

Laser projectors aren't that much different from traditional projectors. With all projectors, something creates light, and that light is then manipulated to create the image on the screen. The only thing changes with laser projectors is *what* is creating the light.

In the purest form, red, green and blue lasers are "defocused" to fill an entire DLP, LCOS, or LCD chip. The lasers don't scan the chip (or the screen). So, in fairness, "laser projector" is about as much of a misnomer as "LED TV." Both terms refer to the of name the technology that creates the light -- lasers and light-emitting diodes, respectively -- not the technology that creates the image as a whole. "DLP projector" and "LCD TV" would be more accurate.

Lasers replace the UHP, Xenon, and other lamps found in current projectors.

Laser projectors are especially accurate due to a fast and stable projection with high repetition rate, so that visual flickering is reduced. Z-LASER laser projectors are optimized for 2D and 3D applications and offer extremely accurate projection in the millimeter range. The optical angles (80 ° x 80 °) allow the coverage of large work areas.

Working Principles of Laser Projector

The laser projector is where the laser beam is controlled and scanned to produce the effects you see in the show.

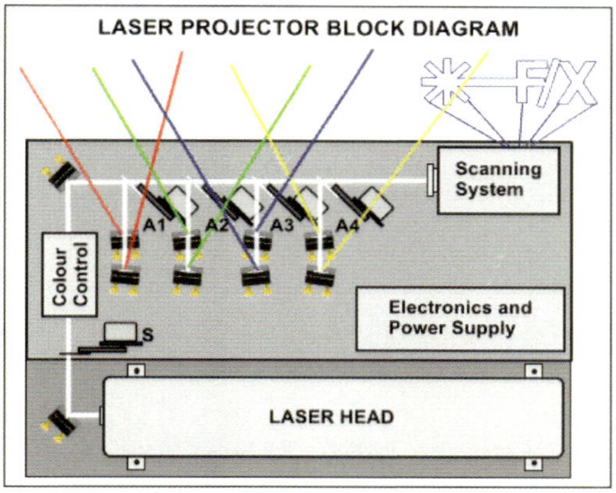

In the block figure above, the laser beam first encounters the shutter (marked S on the diagram). This is typically an actuator or solenoid designed to turn the beam on or off (at slow speeds). In most modern laser projection systems the shutter is gravity or spring- loaded as a safety feature. In the event of a power failure to the laser projector, the beam is blocked by the gravity or spring-loaded shutter so that no projection can take place.

Actuators

One of the most common devices in laser projection equipment is the actuator - used for shutters, colour boxes and in beam tables. An actuator can be thought of as a limited rotation electrical motor that only moves a few degrees when power is applied. In a beam table, the shaft of the actuator is fitted with a lightweight arm that has a mirror glued to the end of it. It is mounted below a laser beam at a 45 degree angle to the beam, such that the mirror is close to, but not touching, the laser beam. Sending an electrical signal to the actuator pulls the arm (and mirror) into the beam so that the laser is deflected at 90 degrees to it's original line of travel and into other optical components.

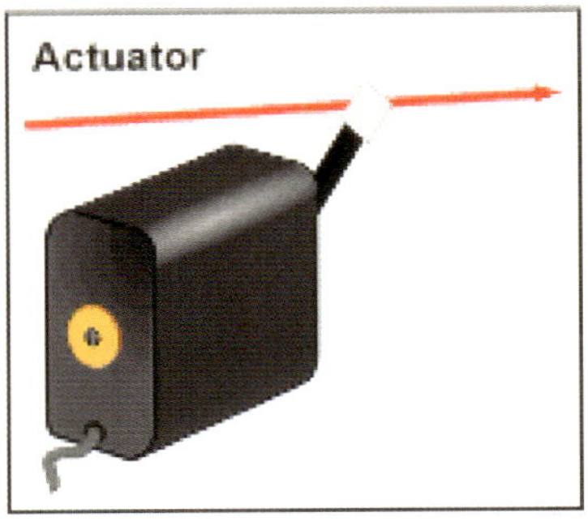

Colour Control

The laser beam next encounters the colour control area. Laser shows either use a tandem laser pair (an Argon laser for the blue and green and a Krypton laser for the red) where the beams from the two lasers are combined to form a "white" beam; or a white light laser which produces red, green and blue from one laser. Colour control can be either subtractive - where unwanted colours are subtracted from the beam; or additive where colours are added to make the desired colour.

Laser Lines

Laserists (and scientists) speak of "lines" (measured in nanometers - nm for short) rather than colours, as the colour of the beam perceived by the eye is a function of the frequency of the photon oscillations. A HeNe laser typically produces a very pure red colour at 632.5 nm. Below is a chart showing some of the most common lines produced by lasers used in light shows.

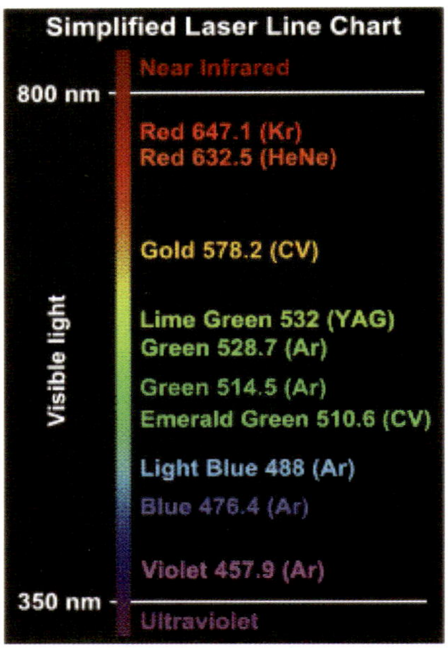

A large and very detailed chart (in.tif format) showing all of the lines emitted by various types of lasers

The most typical forms forms of colour control used in laser shows (when you have a 'white' beam) are a colour box and a PCAOM (Poly Chromatic Acousto Optic Modulator).

Colour Box

The colour box is a subtractive device that uses three actuators with dichroic filters mounted on the ends of the arms (one cyan, one magenta and one yellow dichro) to select one of seven colours that will be displayed. It has a relatively slow response time as it is a mechanical device thus it can only produce a limited number of colour changes in each frame. It has the advantage of low cost and simple implementation. Depending on which colour is desired in the output, one or more of the dichroic filter is inserted into the beam to subtract (block) the unwanted colour(s). The colour box can output white, red, green. blue, cyan, magenta and yellow colours.

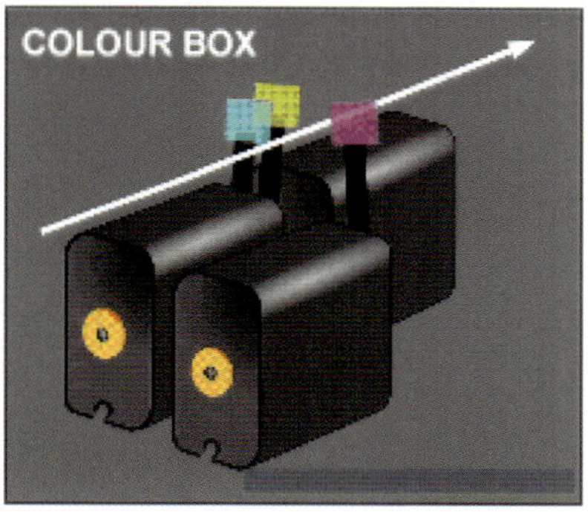

Fig: Colour box showing how different combinations of Dichros produce the colours

PCAOM

At the other end of the scale is the Poly Chromatic Acousto Optic Modulator (PCAOM), a solid state, in-line device that allows for continuous brightness control of multiple laser lines to generate 16.7 million or more colours at MHz speeds. The PCAOM acts as a bulk diffraction grating which gives brightness control over the individual laser lines. Just as a TV set combines variable brightness of Red, Green and Blue to make the different colours you see on the screen, the PCAOM allows for variable brightness control of a number of laser lines (colours) that are added to create the desired output colours for projection. For example, one can combine 100% of the red line(s) with 50% of the green line(s) to produce an orange beam.

A PCAOM consists of a driver card or electronics package that takes the colour control signals and converts then to an RF signal that is fed to the cell. The cell is a crystal with a transducer bonded to it, that is inserted at the correct angle into the laser beam. The PCAOM has the advantages that it can produce 16.7 million (or more) colours, has very high response speed as it is a solid state device thus allowing for multiple colours and blanking within a single frame.

Eight line PCAOM driver card with 3 PCAOM cells to it's right

Beam Table

STP8 Actuator

Beam effects, especially those that are to be directed at bounce mirrors, are usually generated by a beam table. The beam table acts as a digital beam switching device since the actuators can only be in or out of the beam.

The beam table system uses the mechanical movement of the actuator (with it's arm and mirror) to the same position each time, to deflect the beam to optics on kinematic (X-Y adjustable) mounts. These beams are very stable and accurate as they can only be on or off due to the switching nature of the actuator. The kinematic mounts allow the beam to be precisely positioned so as to target remote bounce mirrors or other outboard effects. The beam table area can also contain other optical effects such as lumia wheels and diffraction gratings.

In the projector block above, a four position beam table consisting of actuators A1 to A4 is shown as an example - large professional projectors may have up to 16 beam table positions. When one of the actuators is energized, the mirror on the end of the arm deflects the beam towards the two kinematic mounts positioned behind it. If the first mount is equipped with a 50/50 beamspliter, and the second mount is equipped with a mirror, the beam table position outputs two beams (each half as bright as the input beam). Beam tables used with higher power lasers can have a combination of beam splitters and mirrors to output 4 (quad position) or more beams from each beam table position.

The output beams can be precisely positioned with adjustment screws on the the back of the kinematic mounts, so as to target remote mirrors that can then deflect the beam in a different direction. Using a four position beam table equipped with 50/50 splitters and mirrors, one can generate 8 beams out. By deflecting each of these beams from a remote bounce mirror, it gives the impression that there are many more beams in the venue (16 beams if each beam is bounced once from a remote mirror).

The beam table actuators can also be sequenced to "chase" the beams around the venue. With a high speed sequence (where each actuator is engaged 15-20 times a second), the illusion that the beam table is projecting 8 beams at the same time is created. The beams are being switched on and off too fast for the eye to see thus it appears that all the beams are on at the same time.

Beam Table Colour

Some beam tables use "fixed colour" where the 50/50 beamspliter is replaced by a dichroic filter. When the actuator energizes, the dichroic deflects a beam of a fixed colour while the mirror behind it deflects the remaining colours passed by the dichroic filter. Whenever a particular beam table position is activated, it always outputs the same colours.

In a beam table equipped with 50/50 (or other combinations of) beamspliters, each position can output any colour - all beams from that position will be the same colour). By synchronizing colour changes in the colour control system with the movement of the beam table actuators, each position can display any desired colour that the system can generate.

Scanning System

The most important part of laser projection system are the galvos, more commonly called scanners. Graphics, animations, abstracts and dynamic beam effects are generated by X-Y scanning of the laser beam.

Support Electronics

The projector cabinet (or a separate electronics rack) usually includes all of the power supplies, cooling fans and driver cards used to make the projection system work.

Support electronics may also include safety features such as power-off interlocks and scan fail detectors.

Laser Pointer

The laser pointer is a low cost portable laser that can be carried in the hand. It is designed for use during presentations to point out areas of the slide or picture being presented, replacing a hand held wooden stick or extendable metal pointer. It is superior over older pointers because it can be used from several hundred feet away in a darkened area and because it produces a bright spot of light precisely where the user desires. It has also caught on as an all-purpose pointing tool and has become so commonplace that laws have been passed to restrict its use.

Raw Materials

A laser diode is less complicated than many types of consumer electronic equipment. It consists of a laser diode, a circuit board, a case, optics, and a case. Some of the electrical components on the circuit board and the laser diode are made of semiconductor materials, metals, and ceramics. The semiconductor materials include compounds (materials made of more than one pure element) made of aluminum, gallium, arsenic, phosphorus, indium, and similar elements. These compounds are used in a variety of semiconductor products. Semiconductors also contain metals such as aluminum, gold, and tantalum.

The circuit board is typically made of a resin (plastic) such as epoxy with glass fibers in it to strengthen it. Electricity is conducted to the various components on the circuit board with lines of metal such as aluminum and copper. Individual components placed on the circuit board include diodes, the laser diode, capacitors, and resistors. Semiconductor parts, such as the diodes are encapsulated in plastic with metal leads that are connected to metal pads on the circuit board with solder (a metal alloy traditionally made of tin and lead, but now containing less lead and other metals as substitutes). Non-semiconductor parts, such as resistors and capacitors, are made of a variety of metals, plastics, and ceramics (including glass).

The collimating optics can be glass, but less expensive acrylic plastics are used in most laser pointers. The case can be made of any material, such as metal, plastic, or even wood. It contains metal (usually brass) contacts for the batteries.

Design

The design of the laser pointer depends on the electrical requirements of the laser diode, the desired lifetime of the power supply, and the drive to produce smaller consumer products. The smallest laser pointers are less than two inches in length, but some laser pointers are designed to look like pens. The longer laser pointers can hold AAA or AA batteries, which provide a longer lasting power supply than the watch batteries used in the shorter laser pointers. Most laser pointers use two or three batteries.

Manufacturing Process

The red laser pointer is the most common laser pointer. Other laser pointers use different laser diode assemblies, but are produced in a similar fashion, so the red laser pointer manufacturing process and diagram are used.

Laser Diode

- The laser diode is produced in a semiconductor fab (a factory where semiconductor materials are produced in very clean and carefully controlled conditions). The substrate is the base material on which other materials will be deposited. A wafer of the substrate is produced, cleaned, and prepared. Then it goes through several steps where layers of material are deposited on it. Some of these layers are only several atoms thick. These layers can be conductive (metals such as aluminum and gold) or semiconductors. These layers can also be altered by exposure to other chemicals. After all materials are added to the wafer, it is diced (cut apart, usually into rectangular sections) into individual diodes. The diodes are tested either on the wafer or after separation, and non-functioning ones are scrapped (thrown away). Working laser diodes are then packaged in a plastic container with metal leads for electrical connection.

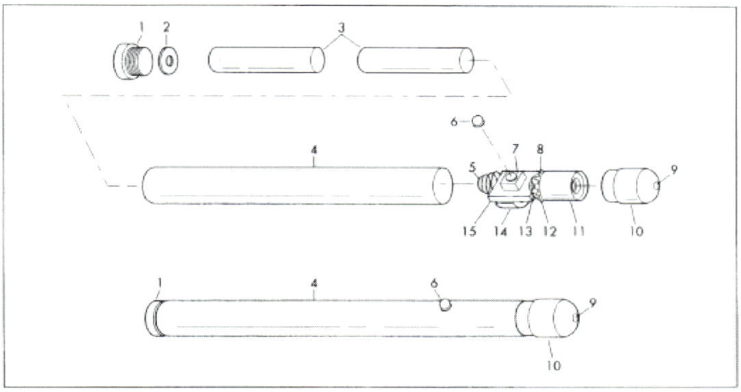

Fig: 1. Cap. 2. Electrically insulative cushion ring. 3. Batteries. 4. Metal barrel. 5. Metal spring. 6. Button. 7. Switch. 8. Spring plate. 9. Laser hole. 10. Metal barrel. 11. Lens barrel. 12. Laser module. 14. Locating frame. 15. Electrically insulative plate

Circuit Board

- The circuit board contains the circuitry that makes the laser pointer function. It contains the switch, the laser diode, and the components of the control circuitry, typically a photodiode, diodes, resistors, and capacitors. These parts are placed on the circuit board, sometimes with an adhesive, and then are soldered in place. Soldering is a process where two metal objects are placed in contact and solder is melted around them so that when it cools, it surrounds both of them and holds them together. Solder is used instead of glue because it sticks to metal and because it conducts heat and electricity.

Collimating Optics

- The collimating optics in a laser pointer consist of a single lens that focuses the cone of light exiting the laser diode into a narrower beam that produces a narrower spot over a longer distance. Plastic lenses are injection molded, a process wherein molten plastic is forced into a mold. The plastic cools and solidifies, then the mold is pulled apart and the lens is removed. It is ground and polished to a smooth surface so that the light from the laser diode will not bounce off of imperfections on the surface.

Laser Diode Assembly

- The laser diode and the collimating optics are put together with a plastic holder to form the laser diode assembly. Most laser diode assemblies have a metal spring attached at the back. This spring makes contact with the batteries in the laser diode and is part of the circuit that draws electricity from the batteries.

Case Construction and Final Assembly

- The case is a tube with space for the laser diode assembly and the batteries. The laser diode assembly is pushed or screwed into one end of the case. The interior of the case is made of brass or has a brass strip (glued or riveted in place) running down the battery space. The battery space end piece also has an exposed brass area or is made of brass. When this end piece is pushed or screwed into the case, it contacts the other side of the batteries to complete the electrical circuit that allows electricity to flow from the batteries to the laser diode assembly.
- The case also has a switch button (a piece of plastic sticking through a hole cut in the side of the case) that must be pushed and held for the laser pointer to work. When this button is pushed, the switch on the circuit board closes, electricity flows from the batteries to the laser pointer assembly, and the laser pointer produces a beam of light.
- After the laser pointer is assembled and tested, a safety label is added. This label describes the rating of the laser in terms of power output, notes which regulations govern its use, and warns the user to avoid direct eye exposure.

Quality Control

A semiconductor manufacturer uses highly controlled processes that have been developed in laboratories and then transferred to the fabrication facility. Laser diodes are tested to make sure that they work after fabrication as well. Each other component is also tested to make sure that it works. Most manufacturing facilities will randomly test their products and use statistical control methods to provide quality products.

When the laser diode assembly or the laser pointer is finally assembled it will be powered and tested with a light detecting device, such as a photodiode, to measure its power output. Laser pointers are Type IIIA laser devices and must produce 5 mW (milliwatt, one thousandth of a watt) of power or less for the United States market. Laser pointers for the European market are typically Class II laser devices and must produce less than 1 mW. These restrictions are for safety purposes.

Brightness and Color

The apparent brightness of the illuminated spot depends strongly on the wavelength of the emitted light. Most devices operate in the red spectral region, where the sensitivity of the eye rapidly decreases with increasing wavelength. Devices with 650-nm output appear about twice as bright as those emitting the same power at 670 nm, and 635-nm devices still about two times brighter. However, the shorter-wavelength laser pointers are typically more expensive. This is particularly true for green lasers, which are significantly brighter than their red counterparts, but are still expensive. They involve a diode-pumped solid-state laser and a frequency doubler. Due to the typically poor conversion efficiency of the frequency doubler at low power levels, hundreds of milliwatts of infrared (typically 1064-nm) light are required for generating a few milliwatts in the green, and the batteries will accordingly not last very long, unless they are comparatively heavy.

Range of a Laser Pointer

Lay persons often ask what is the range of a laser pointer is, and responding to this interest some producers specify some more or less questionable numbers.

If the question is meant to be how far the light of a laser pointer can propagate, the correct answer is that there is no limit, provided that the light is not absorbed or scattered away in the atmosphere. However, the beam area will eventually become larger due to the beam divergence, so that the intensity e.g. on a screen will be reduced even if the overall power remains constant. Accordingly, an airplane pilot looking down into such a beam from an altitude of 10 km will not be disturbed by the remaining small intensity.

The range of a laser pointer may also be understood as the maximum distance from which the spot on the screen can be seen. That kind of range is normally not limited by the beam divergence but by the overall optical power (apart from the wavelength and level of ambient light), since the issue is not the comparatively minor divergence on the way from the laser pointer to the screen, but rather the huge divergence of the scattered light on the way back. Therefore, someone standing next to the illuminated screen would easily see the spot when it is already hardly perceivable from the position of the laser pointer.

In order to use a laser pointer in a safe manner, follow the rules below:

- Never intentionally direct a laser beam towards your eyes or the eyes of others.
- Do not point a laser pointer at shiny/mirror-like surfaces like polished metal or glass. The reflected beam can hit you or someone else in the eye.
- Do not view a laser beam through light-collecting optics such as binoculars and microscopes.
- Ensure that the laser is inactivated (the button is not pressed) when facing an audience during a presentation. It should only be activated when facing the screen or board.
- Do not purchase a laser pointer that has an output power of greater than 5 mw by making sure the label on the pointer says Class 3a (an image of a label is below; sometimes inscribed using Roman numerals: Class IIIa). Some green laser pointers can produce up to 300 mw of power (they are very dangerous Class 3b laser devices)!! As the beam power goes up from 5 mw, the hazard also increases proportionally.
- Purchase a red-beam laser pointer for use in training or meetings since the beam is not perceived to be as bright as the green beams at the same output power. Thus, the red-beam laser pointers are less likely to cause severe flash-blindness and after-images if eye exposures accidentally occur.
- Do not direct a laser beam at a moving vehicle as this can cause an accident due to flash blindness or a startle effect. It is also illegal and you can be arrested and/or fined for it.
- Do not buy laser pointers or laser-related toys for your children. Remove the batteries from your laser pointers if children will have unsupervised access to them.
- Do not purchase laser beam pet toys that are used to entertain people while pets chase the beams around. There is too great a risk of a reflected beam striking the eye of the laser user or somebody else in the room -- at a minimum causing flash-blindness, thus increasing the risk of tripping over obstacles and falling.
- Discourage the inappropriate and irresponsible use of laser pointers.

Applications

A typical use of a hand-held laser pointer is to point at some screen or chart during a presentation, e.g. a conference talk. This is convenient because it can be done from a large distance and requires only a small hand-held device. However, the visibility of the generated spot on the screen is often poor (particularly for red laser pointers with relatively long emission wavelength), and a fast-moving light spot can have a somewhat nervous appearance. Therefore, some people prefer an old-fashioned telescopic pointing device for presentations.

Laser pointers can be useful for, e.g., aligning some machinery, or for certain optical distance measurements.

Laser Scanning

Laser scanning is a non-contact measuring method which uses a laser beam to find the 3D position of a point on an objects surface. By repeating this process several million times the laser scanner is producing a so called "point cloud". This point cloud is a precise capture of a physical object's size and shape and can be imported to a computer for further treatment and visualisation.

There are different types of laser scanners. They can be separated into several types including terrestrial, aerial, mobile, desktop and hand scanners.

Applications for 3D Laser Scanning

- Planning:
  - Save and reliable planning in complex environments;
  - Project time reduction due to precise information;
  - Cost saving due to project time reduction.
- 3D modeling and design:
  - Input for 3D modeling;
  - Crosscheck of new design with as-built situation.
- 2D drafting and documentation:
  - Easy creation of 2D plans and drawings.
  - As-built documentation.
  - Reverse engineering.

- 3D printing
  - Animations and videos.

Uses of 3D Laser Scanning

- Plant industry (e.g. power, chemical, oil, gas plants and facilities);
- Civil industry (e.g. building and infrastructure companies);
- Manufacturing (e.g. factories and ship yards);
- Archaeology (e.g. museums and institutes);
- Heritage and historic preservation (e.g. communities and organisations);
- Law Enforcement (e.g. forensics);
- Art (e.g. artists and museums).

Benefits of 3D Laser Scanning

- Fast
  - Scans with many million points in just a couple of minutes
  - Reduced survey and inspection times
- Accurate
  - Depending on application millimetre or even sub millimetre precision
  - Enables to reduce fabrication tolerances
- Safe
  - Scanning over far distances possible
  - Less exposure to hazardous environments
- Flexible
  - Data can be transferred to almost any CAD system
  - Easy site setup helps flexible scheduling
  - Complicated objects and structures can be measured
- Inexpensive
  - Shorter planning and design time
  - Less rework
  - Less engineering errors due to missing information

Laser Printing

A laser printer is a popular type of personal computer printer that uses a non-impact (keys don't strike the paper), photocopier technology. When a document is sent to the printer, a laser beam "draws" the document on a selenium-coated drum using electrical charges. After the drum is charged, it is rolled in toner, a dry powder type of ink. The toner adheres to the charged image on the drum. The toner is transferred onto a piece of paper and fused to the paper with heat and pressure. After the document is printed, the electrical charge is removed from the drum and the excess toner is collected. Most laser printers print only in monochrome. A color laser printer is up to 10 times more expensive than a monochrome laser printer.

IBM introduced the first laser printer in 1975 for use with its mainframe computers. In 1984, Hewlett-Packard revolutionized laser-printing technology with its first LaserJet, a compact, fast, and reliable printer that personal computer users could afford. Since then, laser printers have decreased further in price and increased in quality. Hewlett Packard continues to be the leading manufacturer with competitors including Lexmark, Oki data, and Xerox.

The laser printer is different from an inkjet printer in a number of ways. The toner or ink in a laser printer is dry. In an inkjet, it is wet. Over time, an inkjet printer is about ten times more expensive to operate than a laser printer because ink needs replenishing more frequently. The printed paper from an inkjet printer will smear if wet, but a laser-printed document will not. Both types of printer operate quietly and allow fonts to be added by using font cartridges or installing soft fonts. If your printing needs are minimal, an inkjet printer is sufficient. But if your printing volume is high, consider buying a laser printer.

Working

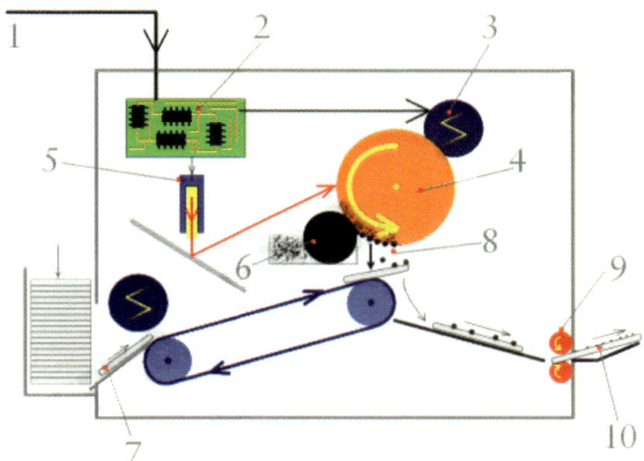

When you print something, your computer sends a vast stream of electronic data (typically a few megabytes or million characters) to your laser printer. An electronic circuit in the printer figures out what all this data means and what it needs to look like on the page. It makes a laser beam scan back and forth across a drum inside the printer, building up a pattern of static electricity. The static electricity attracts onto the page a kind of powdered ink called toner. Finally, as in a photocopier, a fuser unit bonds the toner to the paper.

1. Millions of bytes (characters) of data stream into the printer from your computer.
2. An electronic circuit in the printer (effectively, a small computer in its own right) figures out how to print this data so it looks correct on the page.
3. The electronic circuit activates the corona wire. This is a high-voltage wire that gives a static electric charge to anything nearby.
4. The corona wire charges up the photoreceptor drum so the drum gains a positive charge spread uniformly across its surface.
5. At the same time, the circuit activates the laser to make it draw the image of the page onto the drum. The laser beam doesn't actually move: it bounces off a moving mirror that scans it over the drum. Where the laser beam hits the drum, it erases the positive charge that was there and creates an area of negative charge instead. Gradually, an image of the entire page builds up on the drum: where the page should be white, there are areas with a positive charge; where the page should be black, there are areas of negative charge.
6. An ink roller touching the photoreceptor drum coats it with tiny particles of powdered ink (toner). The toner has been given a positive electrical charge, so it sticks to the parts of the photoreceptor drum that have a negative charge (remember that opposite electrical charges attract in the same way that opposite poles of a magnet attract). No ink is attracted to the parts of the drum that have a positive charge. An inked image of the page builds up on the drum.
7. A sheet of paper from a hopper on the other side of the printer feeds up toward the drum. As it moves along, the paper is given a strong positive electrical charge by another corona wire.
8. When the paper moves near the drum, its positive charge attracts the negatively charged toner particles away from the drum. The image is transferred from the drum onto the paper but, for the moment, the toner particles are just resting lightly on the paper's surface.
9. The inked paper passes through two hot rollers (the fuser unit). The heat and pressure from the rollers fuse the toner particles permanently into the fibers of the paper.
10. The printout emerges from the side of the copier. Thanks to the fuser unit, the paper is still warm. It's literally hot off the press.

To buying a laser printer, these are some important features to consider:

- Print capacity and speed

Personal laser printers are sufficient for printing an average of 200 pages per week. These are low-end and cost $200 and up. They can print up to eight ppm (pages per minute). A workgroup printer is needed if an average of 1000 pages per week is needed. These print up to 24 ppm and cost $1000 to $6000 and more. Production printers are needed for printing 50,000 or more pages per week. These are quite expensive and are used by commercial publishers. They can print up to 700 ppm and cost $100,000 and up. They can print 24 hours a day, seven days a week.

- Resolution

The standard resolution in most laser printers today is 600 dots-per-inch (dpi). This resolution is sufficient for normal everyday printing including small desktop publishing jobs. A high-end production printer might have a resolution of 2400 dpi. Some laser printers still use a resolution of 300 dpi. This resolution can cause jagged lines to appear on the outer edge of an image. Hewlett Packard created RET (Resolution Enhancement Technology) to correct this. RET inserts smaller dots at the edges of lines and to smooth the rough edges. RET does not improve the resolution, but the document looks better. If you purchase a printer with 300 dpi, make sure it has RET.

- Printer languages

Printer Control Language (PCL) is the standard printer language for Hewlett Packard and most other laser printers (which are mostly HP-compatible). PCL is used for printing letters, database printouts, spreadsheets, and simple graphics. Postscript printers are used with desktop publishing software and drawing packages. Postscript printers are the norm for Apple Macintosh printers. A laser printer that comes with Postscript installed is more expensive. A laser printer that uses PCL can be upgraded to Postscript by installing a software driver provided by the manufacturer of the laser printer. The printer might require more memory when upgraded to use Postscript. This is because a laser printer needs the entire image in memory before printing, and a Postscript printer requires more memory to process than a PCL printer does. The application being used must support Postscript in order for the laser printer to print Postscript documents.

- Paper handling

Paper handling is important when shopping for a laser printer. Most laser printers use letter-size, cut-sheet paper. High-end production printers use continuous sheet paper. Laser printers can print on transparencies, adhesive labels, and lightweight cards. A laser printer with duplex printing can print on one side of the paper, turn the paper over, and print on the other side. Most laser printers, however, use simple printing with manual duplex printing. Manual duplex printing is achieved by changing the print options in the printer's properties or printing one side and taking that same paper and reinserting it into the printer to print on the other side.

- FPOT and warm-up time

A final consideration in purchasing a printer is FPOT (first paper out time) and warm-up time. When a laser printer receives data from the computer to print, it takes 5 to 30 seconds to prepare the printer to print a new job. This is in addition to the time it takes to actually print the document. The warm-up time is as important. When the printer is turned on, it needs time to warm up the fuser to operating temperature. If the printer has a standby mode or is turned off between printing jobs, the warm-up time becomes even more important. Large workgroup and production printers can take 5 to 15 minutes to warm up. This waiting period can hinder overall productivity.

Audience Scanning

Audience scanning is one of the most beautiful effects that can be created with a laser show projector, but it is also one of the most dangerous audience scanning is where laser light is intentionally projected directly into an audience area, allowing people to interact and actually reach out and touch the light. While this effect is perhaps the most beautiful that can be produced with a laser, it also carries the greatest risk, and requires that the laser operator take certain safety precautions to ensure that the light level does not exceed recognized safety standards. In addition, there are certain regulations surrounding the performance of audience scanning in different geographic territories around the world.

Safety Hazards

The following diagram provides a great illustration and starting point, to help explain why directly viewing a laser beam can be potentially hazardous.

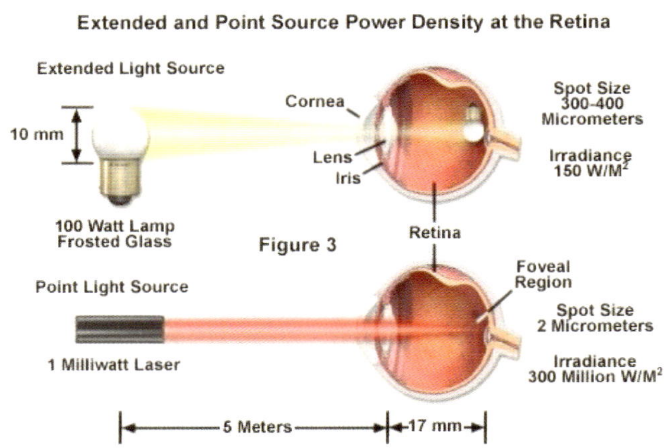

As the figure above helps illustrate, the human eye is able to focus laser light much more efficiently than it can other types of light (such as sunlight). Better focusing leads to better concentration of the light. And, just like when holding a magnifying glass to focus a beam of sunlight into a given area, the human eye will focus laser light into a very small spot. At very high levels, the concentrated energy could become hazardous to a person's eyes.

This is a major reason why there are regulations surrounding the performance of audience scanning, especially in the United States, as well as in some other countries.

Audience scanning immerses the audience with light in many shapes and sizes (laser beams, cones, fans, waves, sheets, etc) so that the laser light is actually touching people. It is a truly an immersive effect that allows people to virtually touch and interact with the light. Laser lighting can be shined directly at the audience, reflect off natural and artificial structures in the environment, or focused on the stage and performer. Regardless of the directionality of the display, there is a risk for over exposure to the human eye.

Medical Uses of Lasers

Lasers can be used for many medical purposes the number of uses for medical lasers multiplied over the years. Among other applications, these include cutting into tissue in surgical procedures; reshaping the cornea of the eye to improve sight; cleaning clogged arteries; burning away cavities and whitening the teeth; removing unwanted hair, wrinkles, birthmarks, and freckles; and reshaping the face in plastic surgery procedures.

The Advent of the Laser Scalpel

Early experimenters with medical lasers pointed out that there are surgical operations that are difficult to perform with the conventional scalpel and that a laser beam might be used instead. Initial trials showed that a finely focused beam from a carbon dioxide gas laser could cut through human tissue easily and neatly. The surgeon could direct the beam from any angle by using a mirror mounted on a movable metal arm.

Several advantages of laser surgery quickly became apparent. First, the light beam is consistent, which means that it gives off the same amount of energy from one second to the next. So as long as the beam is moving along, the cut it makes (the incision) does not vary in depth; whereas when using a scalpel a doctor can accidentally make part of the incision too deep. A second advantage of the surgical laser is that the hot beam cauterizes, or seals off, the open blood vessels as it moves along. (This works well mainly for small vessels, such as those in the skin. The doctor still has to seal off the larger blood vessels using conventional methods.) Still another advantage is that the cells in human tissue do not conduct heat very well, so the skin or any other tissue near the laser incision does not get very hot and is not affected by the beam. This advantage of laser surgery is very helpful when a doctor must operate on a tiny area that is surrounded by healthy tissue or organs.

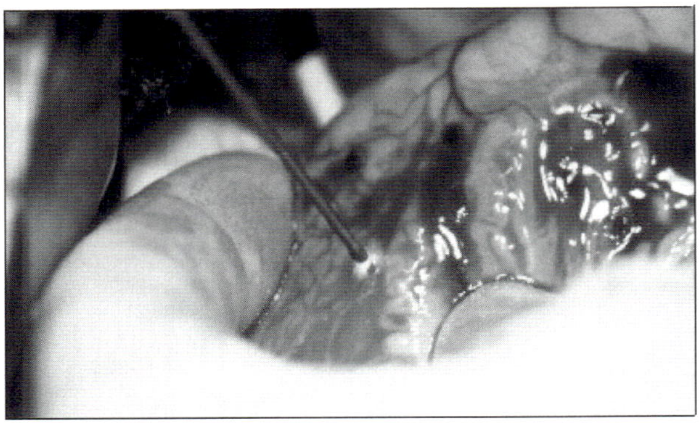

In this photo taken during open-heart surgery, a doctor uses a laser probe to punch small holes in the patient's heart muscle to increase the organ's blood flow

It should be pointed out that the "laser scalpel" is not necessarily the best tool to use in every operation. Some doctors feel that while the laser is useful in some situations, it will never totally replace the scalpel. Others are more optimistic and see a day when more advanced lasers will make the scalpel a thing of the past.

The second of these views may prove to be the most accurate, for surgical use of lasers is rapidly advancing. At first, lasers were considered most effective in operating on areas that are easy to reach—areas on the body's exterior, including the skin, mouth, nose, ears, and eyes. But in recent years doctors have demonstrated remarkable progress in developing laser techniques for use in internal exploration and surgery. Of course, in order to be able to direct the laser beam the doctor must be able to see inside the body. In some cases this is a simple matter of making an incision and opening up the area to be operated on. But there are situations in which this step can be avoided.

Cleaning Arteries with Light

For instance, lasers are increasingly used to clean plaque from people's arteries. Plaque is a tough fatty substance that can build up on the inside walls of the arteries. Eventually the vessels can get so clogged that blood does not flow normally, and the result can be a heart attack or stroke, both of which are serious and sometimes fatal. The traditional method for removing the plaque involves opening the chest and making several incisions, a long and sometimes risky operation. It is also expensive and requires weeks for recovery.

An effective alternative is to use a laser beam to burn away the plaque. The key to making this work is the doctor's ability to see inside the artery and direct the beam, another area in which fiber optics and lasers are combined into a modern wonder tool. An optic fiber that has been connected to a tiny television camera can be inserted into an artery. These elements now become a miniature sensor that allows the doctor and nurses to see inside the artery while a second fiber is inserted to carry the bursts of light that will burn away the plaque.

The technique works in the following way. The fiber-optic array is inserted into a blood vessel in an arm or leg and moved slowly into the area of the heart and blocked arteries. When the array is in place the laser is fired and the plaque destroyed, and then the exhaust vapors are sucked back through a tiny hollow tube that is inserted along with the optical fibers. When the artery has been cleaned out the doctor removes the fibers and tube, and the operation is finished. This medical process is known as laser angioplasty. It has several obvious advantages. First, no incision is needed (except for the small one in the vessel to insert the fibers). There is also little or no bleeding, and the patient can enjoy total recovery in a day or two.

Laser angioplasty does have some potential risks that must be considered. First, when the laser beam fires at the plaque it must be aimed very carefully because a slight miss could cut through the wall of the artery and cause serious bleeding. The patient's chest would then have to be opened up after all. Another problem involves small pieces of burnt debris from the destroyed plaque. If these enter the bloodstream, they can cause blockage in smaller blood vessels, bringing further complications. Fortunately, continuous technical advancements have considerably reduced these risks, and the number of successful laser angioplasties performed is increasing each year.

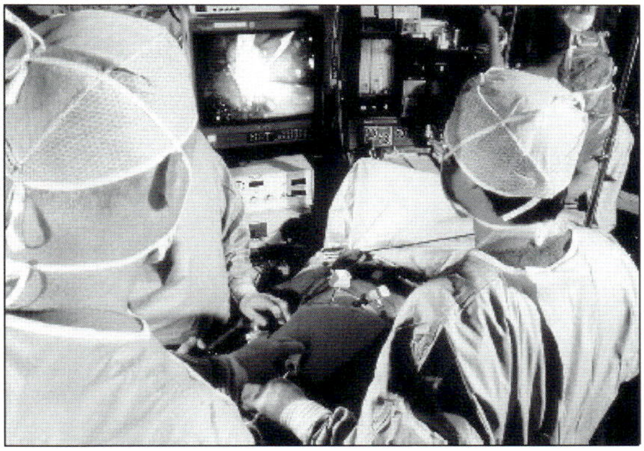

Surgeons use a tiny laser to cut away tissue in a gallbladder operation. The laser and a tiny camera are inserted into the navel, so no abdominal incision is necessary

Lasers Heal and Reshape the Eyes

Some of the most remarkable breakthroughs for medical lasers have been in the area of ophthalmology, the study of the structure and diseases of the eye. One reason that laser beams are so useful in treating the eye is that the cornea, the coating that covers the eyeball and admits light into the interior of the eye, is transparent. Since it is designed to admit ordinary light, the cornea lets in laser light just as well and remains unaffected by the beam.

First, the laser is very useful in removing extraneous blood vessels that can form on the retina—the thin, light-sensitive membrane at the back of the eyeball. It is on the retina that the images of the things the eye sees are formed. Damage to the retina can sometimes cause blindness, the most common form in the United States resulting from diabetes (a disease characterized by high levels of blood sugar) when, in some advanced cases, hundreds of tiny extra blood vessels form on the retina. These block light from the surface of the membrane, resulting in partial or total blindness.

The laser most often used in the treatment of this condition is powered by a medium of argon gas. The doctor aims the beam through the cornea and burns away the tangle of blood vessels covering the retina. The procedure takes only a few minutes and can be done in the doctor's office. The laser can also repair a detached retina—one that has broken loose from the rear part of the eyeball. Before the advent of lasers detached retinas had to be repaired by hand, and because the retina is so delicate this was a very difficult operation to perform. Using the argon laser, the doctor can actually "weld" the torn retina back in place. It is perhaps a strange coincidence that Gordon Gould, one of the original inventors of the laser, later had one of his own retinas repaired this way.

Another condition that affects the eye is glaucoma, which is characterized by the build-up of fluid in the eye. Normally the eye's natural fluids drain away a little at a time, and the eye stays healthy. In eyes impaired with glaucoma the fluid does not drain properly, and the buildup affects vision; blindness can sometimes result. In some cases drugs can be used to treat glaucoma. If the drugs fail, however, many doctors now turn to the laser to avoid conventional surgery. The laser punches a hole in a preplanned spot and the fluid drains out through the hole. Again, the treatment can be performed in a doctor's office instead of a hospital.

Using Lasers for Eye Surgery

The laser works like a sewing machine to repair a detached retina, the membrane that lines the interior of the eye. The laser beam is adjusted so that it can pass harmlessly through the lens and focus on tiny spots around the damaged area of the retina. When it is focused, the beam has the intensity to "weld" or seal the detached area of the retina back against the wall of the eyeball.

Perhaps most exciting of all the eye-related laser applications is the reshaping of the eye's cornea, a technique widely known as LASIK (which stands for *L*aser-*A*ssisted). As Breck Hitz describes it:

> The patient's eyeglass prescription is literally carved inside the cornea with the beam of an excimer laser [a laser device that produces pulses of ultraviolet, or UV, light]. A small flap of the cornea is first removed with a precision knife and an inner portion of the cornea is exposed to the excimer laser. After the prescription is carved, the corneal flap that was opened is then put back into place over the ablated [surgically altered] cornea.

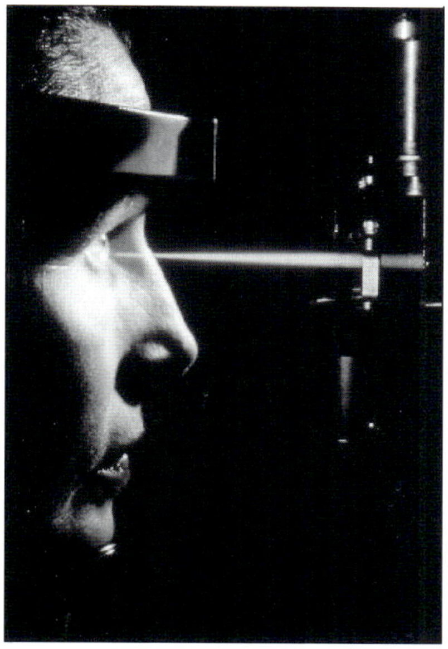

A patient undergoes eye surgery performed by a laser beam.
In addition to treating detached retinas, lasers can remove cataracts

LASIK does not come without risks. The changes it makes in the cornea are permanent, and the danger of unexpected damage is ever present. However, the procedure has become increasingly popular each year; about a million Americans had it done in the year 2000, and about four thousand surgeons in the United States were trained to perform it.

Some Cosmetic uses of Lasers

Medical lasers are also widely used for various types of cosmetic surgery, including the removal of certain kinds of birthmarks. Port-wine stains, reddish purple skin blotches that appear on about three out of every one thousand children, are an example. Such stains can mark any part of the body but are most commonly found on the face and neck.

The medical laser is able to remove a port-wine stain for the same reason that a military laser is able to flash a message to a submerged submarine. Both lasers take advantage of the monochromatic quality of laser light, that is, its ability to shine in one specific color. The stain is made up of thousands of tiny malformed blood vessels that have a definite reddish purple color. This color very strongly absorbs a certain shade of green light. In fact, that is why the stain looks red. It absorbs the green and other colors in white light but reflects the red back to people's eyes.

To treat the stain, the doctor runs a wide low-power beam of green light across the discolored area. The mass of blood vessels in the stain absorbs the energetic laser light and becomes so hot that it is actually burned away. The surrounding skin is a different color than the stain, so that skin absorbs only small amounts of the beam and remains unburned. (Of course, the burned areas must heal, and during this process some minor scarring sometimes occurs.)

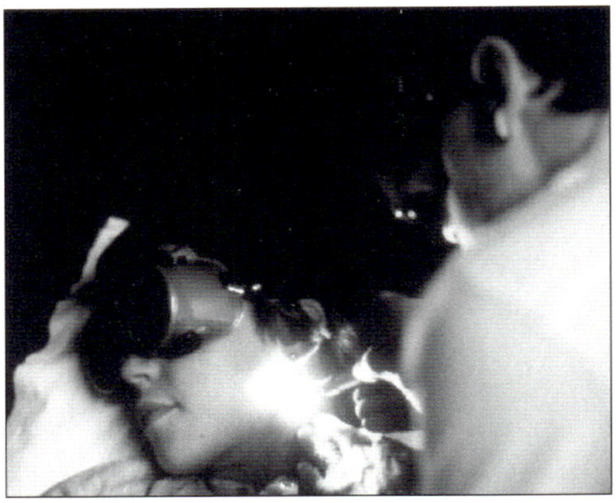

A doctor uses an argon laser to remove a port-wine stain, a kind of birthmark. Unwanted tissue is burned away while normal skin remains undamaged

A similar method is often successful in removing tattoos. A tattoo is formed when very strong dyes are injected with needles into a person's skin. Someone who has been tattooed may decide later in life that he or she does not want the tattoo anymore; and in the past, the only way to remove these designs involved surgery or burning off the tattoo with acid. Luckily, the laser offers an alternative to such extreme measures. The beam bleaches the dyes in the tattoo without burning the surrounding skin. (As in the case of port-wine stains, some light scarring is possible.)

Still another example of a laser-assisted cosmetic procedure is the removal of unwanted hair. "The laser emits a gentle beam of light," explains Chicago cosmetic surgeon Jeffrey Melton, which only is absorbed in the hair follicle (leaving the skin unharmed). The laser delivers energy which is absorbed in the hair and transformed to heat. The heat destroys the hair follicle within a small fraction of a second laser hair removal is good for both facial hair removal and for body hair removal perhaps the most commonly treated areas are facial hair in women.

Laser-assisted Dentistry

Dentistry is another branch of medicine that has benefited tremendously from laser technology. Indeed, lasers have made some people stop dreading a visit to the dentist. No one enjoys having a cavity drilled, of course. It usually requires an anesthetic (a painkiller like novocaine) that causes uncomfortable numbness in the mouth; also, the sound of the drill can be irritating or even sickening to some people.

Many dentists now employ an Nd- YAG laser (which uses a crystal for its lasing medium) instead of a drill for most cavities. The laser treatment takes advantage of the simple fact that the material that forms in a cavity is much softer than the enamel (the hard part of a tooth). The laser is set at a power that is just strong enough to eliminate the decayed tissue but not strong enough to harm the enamel. When treating a very deep cavity bleeding sometimes occurs, and the laser beam often seals off blood vessels and stops the bleeding.

The most often asked question about treating cavities with lasers is: Does it hurt? The answer is no. Each burst of laser light from a dental laser lasts only thirty-trillionths of a second, much faster than the amount of time a nerve takes to trigger pain. In other words, the beam would have to last 100 million times longer in order to cause any discomfort. So this sort of treatment requires no anesthetic.

There are literally hundreds of other medical uses for the laser. Still, numerous medical conditions cannot be helped by laser light. And even in those that do respond to laser treatments, a doctor may have a good reason for choosing a different method in a specific case. The plain fact is that, while the laser is a marvelous medical tool, it cannot cure every ill. Yet the world has seen probably only a small fraction of the laser's potential. After all, this super tool has only existed since 1960, and, considering the medical advances it already has created, the future appears promising indeed.

References

- What-is-3d-scanning: laserdesign.com, Retrieved 26 April 2018
- 3d-scanning-technologies-and-the-3d-scanning-process: aniwaa.com, Retrieved 16 May 2018
- Why-lasers-are-the-future-of-projectors: cnet.com, Retrieved 31 March 2018
- Laser-printer: whatis.techtarget.com, Retrieved 11 July 2018
- Audience-scanning-lasers: lasershowprojector.com, Retrieved 30 June 2018
- Audience-scanning-laser-safety, applications: intl-lighttech.com, Retrieved 10 July 2018
- Medical-Uses-of-Lasers: scienceclarified.com, Retrieved 12 May 2018

7

Laser used in Machining

Machining is a set of processes in which a raw material is cut into a specific shape and size through a controlled material-removal process. The various applications of lasers in machining have been explored in this chapter, such as laser beam machining, laser beam welding, hybrid laser welding, laser shock peening, laser engraving, etc.

Laser Machining

LM is a method of cutting metal or refractory materials by melting and vaporizing the material with an intense beam of light from a laser. Drilling by laser, although costly in energy since material must be melted and vaporized to be removed, is used to cut small holes (0.005 to 0.05 inch [0.13 to 1.3 millimetres]) in materials that are too difficult to machine by traditional methods. A common application is the laser drilling of diamonds to be used as dies for drawing wire. Lasers also are used to drill and cut ceramics and substrates for integrated circuits; the aircraft industry uses CNC-controlled lasers to cut profiles and to drill holes in engine parts.

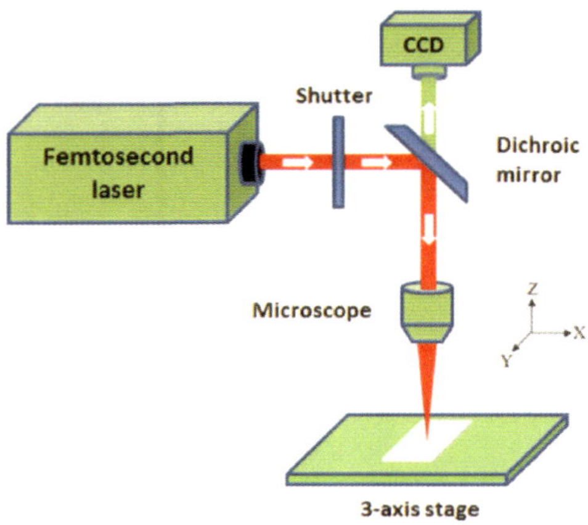

Laser Beam Machining

Laser Beam Machining or more broadly laser material processing deals with machining and material processing like heat treatment, alloying, cladding, sheet metal bending etc. Such processing is carried out utilizing the energy of coherent photons or laser beam, which is mostly converted into thermal energy upon interaction with most of the materials. Nowadays, laser is also finding application in regenerative machining or rapid prototyping as in processes like stereo-lithography, selective laser sintering etc.

Laser beam can very easily be focused using optical lenses as their wavelength ranges from half micron to around 70 microns. Focussed laser beam as indicated earlier can have power density in excess of 1 MW/mm^2. As laser interacts with the material, the energy of the photon is absorbed by the work material leading to rapid substantial rise in local temperature. This in turn results in melting and vaporisation of the work material and finally material removal.

Laser Beam Machining – The Lasing Process

Lasing process describes the basic operation of laser, i.e. generation of coherent (both temporal and spatial) beam of light by "light amplification" using "stimulated emission".

In the model of atom, negatively charged electrons rotate around the positively charged nucleus in some specified orbital paths. The geometry and radii of such orbital paths depend on a variety of parameters like number of electrons, presence of neighbouring atoms and their electron structure, presence of electromagnetic field etc. Each of the orbital electrons is associated with unique energy levels. At absolute zero temperature an atom is considered to be at ground level, when all the electrons occupy their respective lowest potential energy. The electrons at ground state can be excited to higher state of energy by absorbing energy form external sources like increase in electronic vibration at elevated temperature, through chemical reaction as well as via absorbing energy of the photon. Above fig. depicts schematically the absorption of a photon by an electron. The electron moves from a lower energy level to a higher energy level.

On reaching the higher energy level, the electron reaches an unstable energy band. And it comes back to its ground state within a very small time by releasing a photon. This is called spontaneous emission. The same is shown in figures below the spontaneously emitted photon would have the same frequency as that of the "exciting" photon.

Sometimes such change of energy state puts the electrons in a meta-stable energy band. Instead of coming back to its ground state immediately (within tens of ns) it stays at the elevated energy state for micro to milliseconds. In a material, if more number of electrons can be somehow pumped to the higher meta-stable energy state as compared to number of atoms at ground state, then it is called "population inversion". Such electrons, at higher energy meta-stable state, can return to the ground state in the form of an avalanche provided stimulated by a photon of suitable frequency or energy. This is called stimulated emission. Fig below shows one such higher state electron in meta-stable orbit. If it is stimulated by a photon of suitable energy then the electron will come down to the lower energy state and in turn one original photon, another emitted photon by stimulation having some temporal and spatial phase would be available. In this way coherent laser beam can be produced.

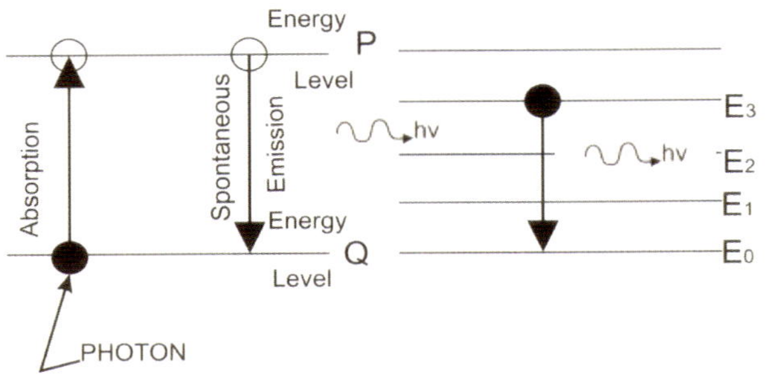

Fig: Energy bands in materials

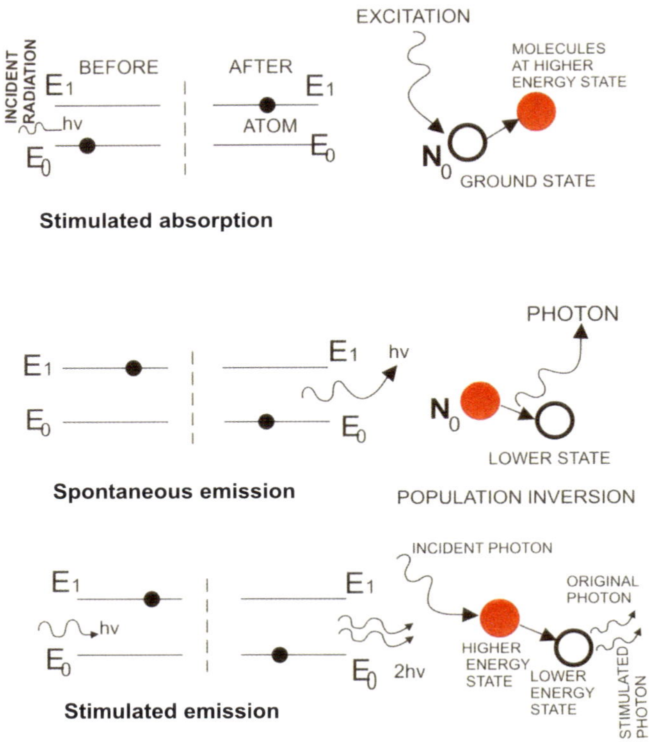

Fig: Spontaneous and stimulated emissions

Figure below shows working of a laser. There is a gas in a cylindrical glass vessel. This gas is called the lasing medium. One end of the glass is blocked with a 100% reflective mirror and the other end is having a partially reflective mirror. Population inversion can be carried out by exciting the gas atoms or molecules by pumping it with flash lamps. Then stimulated emission would initiate lasing action. Stimulated emission of photons could be in all directions. Most of the stimulated photons, not along the longitudinal direction would be lost and generate waste heat. The photons in the longitudinal direction would form coherent, highly directional, intense laser beam.

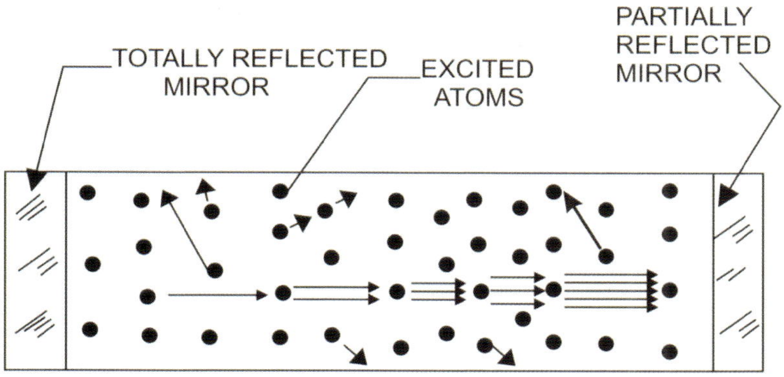

Fig: Lasing action

Lasing Medium

Many materials can be used as the heart of the laser. Depending on the lasing medium lasers are classified as solid state and gas laser. Solid-state lasers are commonly of the following type:

- Ruby which is a chromium– alumina alloy having a wavelength of 0.7 μm.
- Nd- glass lasers having a wavelength of 1.64 μm.
- Nd- YAG laser having a wavelength of 1.06 μm.

These solid-state lasers are generally used in material processing.

The generally used gas lasers are:

- Helium – Neon;
- Argon;
- CO_2 etc.

Lasers can be operated in continuous mode or pulsed mode. Typically CO2 gas laser is operated in continuous mode and Nd – YAG laser is operated in pulsed mode.

Laser Construction

Figure below shows a typical Nd-YAG laser. Nd-YAG laser is pumped using flash tube. Flash tubes can be helical, as shown in Figure below, or they can be flat. Typically the lasing material is at the focal plane of the flash tube. Though helical flash tubes provide better pumping, they are difficult to maintain.

Figure above shows the electrical circuit for operation of a solid-state laser. The flash tube is operated in pulsed mode by charging and discharging of the capacitor. Thus the pulse on time is decided by the resistance on the flash tube side and pulse off time is decided by the charging resistance. There is also a high voltage switching supply for initiation of pulses.

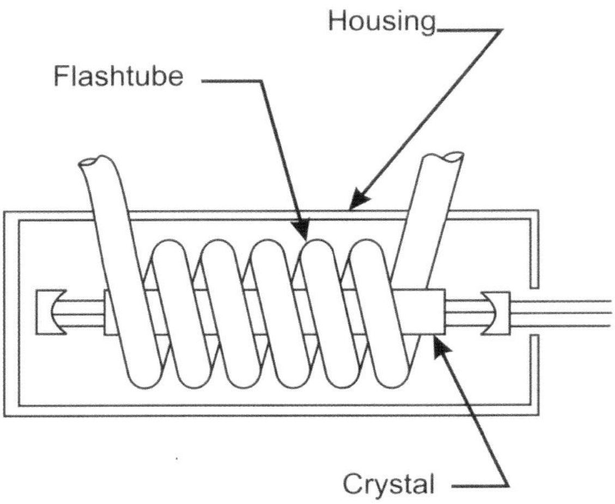

Fig: Solid-state laser with its optical pumping unit

Figure below shows a CO_2 laser. Gas lasers can be axial flow, as shown in Figure below, transverse flow and folded axial flow as shown in figure below, The power of a CO_2 laser is typically around 100 Watt per metre of tube length. Thus to make a high power laser, a rather long tube is required which is quite inconvenient. For optimal use of floor space, high-powered CO_2 lasers are made of folded design.

In a CO_2 laser, a mixture of CO_2, N2 and He continuously circulate through the gas tube. Such continuous recirculation of gas is done to minimize.

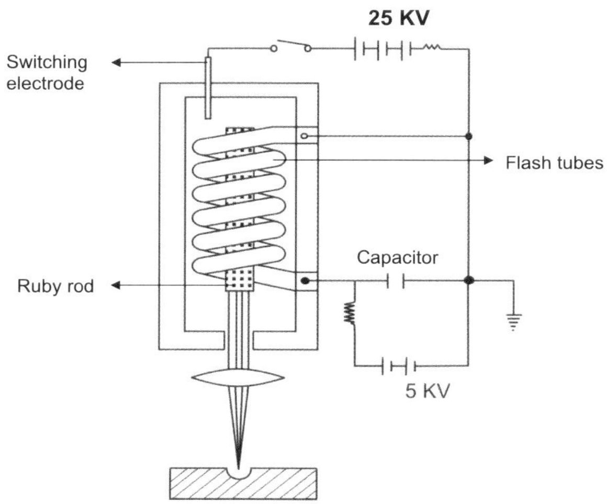

Fig: Working of a solid-state laser

Consumption of gases. CO_2 acts as the main lasing medium whereas nitrogen helps in sustaining the gas plasma. Helium on the other hand helps in cooling the gases.

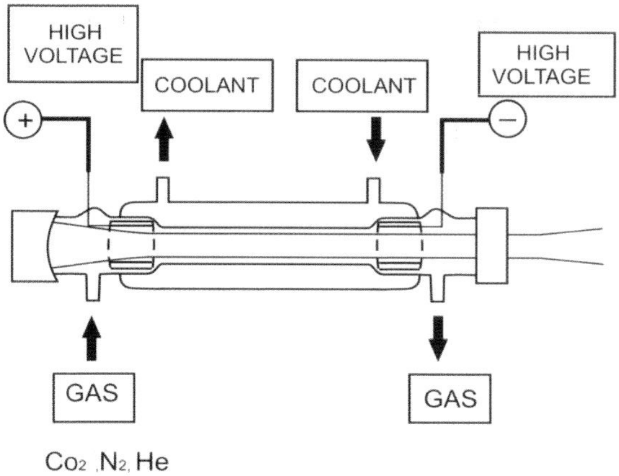

Fig: Construction of a CO_2 laser

As shown in Figure above high voltage is applied at the two ends leading to discharge and formation of gas plasma. Energy of this discharge leads to population inversion and lasing action. At the two ends of the laser we have one 100% reflector and one partial reflector. The 100% reflector redirects the photons inside the gas tube and partial reflector allows a part of the laser beam to be issued so that the same can be used for material processing. Typically the laser tube is cooled externally as well.

As had been indicated earlier CO_2 lasers are folded to achieve high power. Figure below shows a similar folded axial flow laser. In folded laser there would be a few 100% reflective turning mirrors for manoeuvring the laser beam from gas supply as well as high voltage supply as shown in Figure below.

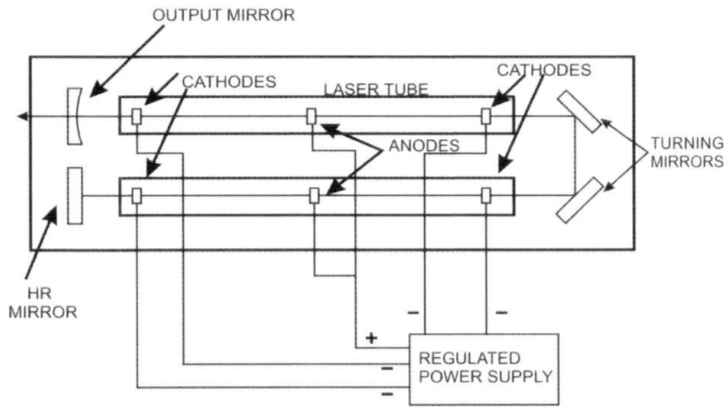

Fig: Construction of folded gas laser

Table below shows the capability and characteristics of common lasers.

Application	Type of laser
Large holes upto 1.5 mm dia. Large holes (trepanned) Small holes > 0.25 mm dia. Drilling (punching or percussion)	Ruby, Nd-glass, Nd-YAG Nd-YAG, CO_2 Ruby, Nd-glass, Nd-YAG Nd-YAG, Ruby

Thick cutting Thin slitting of metals Thin slitting of plastics	CO_2 with gas assist Nd-YAG CO_2
Plastics Metals Organics, Non-metal Ceramics	CO_2 Nd-YAG, ruby, Nd-glass Pulsed CO_2 Pulsed CO_2, Nd-YAG

Table below process characteristics of different lasers:

Lasing materials	Ruby	Nd-YAG	Nd-glass	CO_2
Type	Solid state	Solid state	Solid state	Gas
Composition	0.03 – 0.7% Nd in $Al3O_2$	1% Nd doped Yttrium – Aluminium- Garnet	2-6% Nd in glass	CO_2+He+N_2 (3:8:4)
Wavelength (radiation)	0.69 µm	1.064 µm	1.064 µm	10.6 µm
Efficiency	1% max.	2%	2%	10-15%
Beam mode	Pulsed or CW	Pulsed or CW	Pulsed	Pulsed or CW
Spot size	0.015 mm	0.015 mm	0.025 mm	0.075 mm
Pulse repetition rate (normal operation).	1-10 pps	1-300 pps or CW	1-3 pps	CW
Beam output	10-100 W	10-1000 W	10 – 100 W	0.1 – 10 kW
Peak power	200 kW	400 kW	200 kW	100 kW

Application

Laser can be used in wide range of manufacturing applications:

- Material removal – drilling, cutting and tre-panning
- Welding
- Cladding
- Alloying.

Drilling micro-sized holes using laser in difficult – to – machine materials is the most dominant application in industry. In laser drilling the laser beam is focused over the desired spot size. For thin sheets pulse laser can be used. For thicker ones continuous laser may be used.

Advantages

- In laser machining there is no physical tool. Thus no machining force or wear of the tool takes place;
- Large aspect ratio in laser drilling can be achieved along with acceptable accuracy or dimension, form or location;

- Micro-holes can be drilled in difficult – to – machine materials;
- Though laser processing is a thermal processing but heat affected zone specially in pulse laser processing is not very significant due to shorter pulse duration.

Limitations

- High initial capital cost;
- High maintenance cost;
- Not very efficient process;
- Presence of Heat Affected Zone – specially in gas assist CO_2 laser cutting;
- Thermal process – not suitable for heat sensitive materials like aluminium glass fibre laminate.

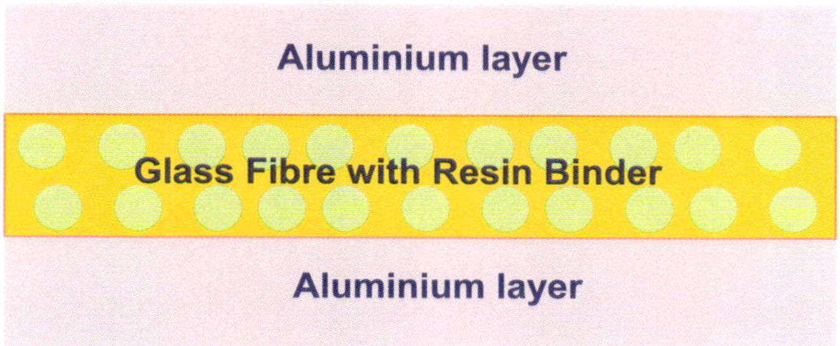

Fig: Aluminium Glass Fibre Laminate – heat sensitive glass fibre layer due to presence of resin as binder

Laser Beam Welding

Laser beam welding is a fusion welding process in which two metal pieces are joined together by the use of laser. The laser beams are focused to the cavity between the two metal pieces to be joined. The laser beams have enough energy and when it strikes the metal pieces produces heat that melts the material from the two metal pieces and fills the cavity. After cooling a strong weld is formed between the two pieces.

Working Principle

It works on the principle that when electrons of an atom gets excited by absorbing some energy. And then after some time when it returns back to its ground state, it emits a photon of light. The concentration of this emitted photon increased by stimulated emission of radiation and we get a high energy concentrated laser beam.

Light amplification by stimulated emission of radiation is called laser.

Main Parts

The main parts or equipment of laser beam welding are:

1. Laser machine: It is a machine that is used to produce laser for welding. The main components of laser machine are shown below.
2. Power source: A high voltage power source is applied across the laser machine to produce laser beam.
3. CAM: It is a computer aided manufacturing in which the laser machine is integrated with the computers to perform welding process. All the controlling action during the welding process by laser is done by CAM. It speeds up the welding process to a greater extent.
4. CAD: It is called as Computer aided Design. It is used to design the job for welding. Here computers are used to design the work piece and how the welding is performed on it.
5. Shielding Gas: A shielding gas may be used during the welding process in order to prevent the w/p from oxidation.

Types of Laser used

1. Gas lasers: It uses mixtures of gases as lasing medium to produce laser. Mixtures of gases such as nitrogen, helium and co_2 are used as lasing medium.
2. Solid-state laser: it uses several solid media such as synthetic ruby crystal (chromium in aluminum oxide), neodymium in glass (Nd: glass), and neodymium in yttrium aluminum garnet (Nd-YAG, most commonly used).
3. Fiber laser: The lasing medium in this type of laser is optical fiber itself.

Characteristics of Laser Beam Welding

1. The power density of laser beam welding is high. It is of the order 1 MW/cm^2. Because of this high energy density, it has small heat-affected zones. The rate of heating and cooling is high.
2. The laser beams produced are coherent (having same phase) and monochromatic (i.e. having same wavelength).
3. It is used to weld smaller sizes spot but the spot sizes can vary from 2 mm to 13 mm.
4. The depth of penetration of the LBW depends upon the amount of power supply and location of the focal point. It is proportional the amount of power supply. When the focal point is kept slightly below the surface of the work piece, the depth of penetration is maximized.

5. Pulsed or continuous laser beams are used for welding. Thin materials are weld by using millisecond-pulses and continuous laser beams are used for deep welds.
6. It is versatile process because it is capable of welding carbon steels, stainless steel, HSLA Steels, aluminum and titanium. Due to high cooling rate, the problem of cracking is there when welding high-carbon steels.
7. It produces high quality weld.
8. This welding process is most popular in automotive industry.

Working

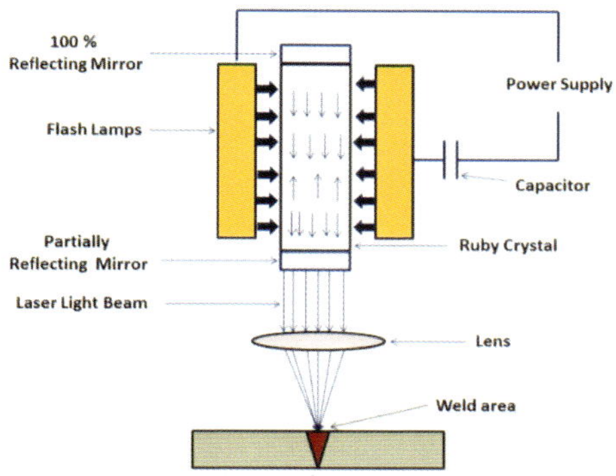

- First the setup of welding machine at the desired location (in between the two metal pieces to be joined) is done.
- After setup, a high voltage power supply is applied on the laser machine. This starts the flash lamps of the machine and it emits light photons. The energy of the light photon is absorbed by the atoms of ruby crystal and electrons get excited to their higher energy level. When they return back to their ground state (lower Energy state) they emit a photon of light. This light photon again stimulates the excited electrons of the atom and produces two photons. This process keeps continue and we get a concentrated laser beam.
- This high concentrated laser beam is focused to the desired location for the welding of the multiple pieces together. Lens are used to focus the laser to the area where welding is needed. CAM is used to control the motion of the laser and work piece table during the welding process.
- As the laser beam strikes the cavity between the two metal pieces to be joined, it melts the base metal from both the pieces and fuses them together. After solidification we get a strong weld.
- This is how a laser beam welding works.

Advantages

- It produces high weld quality.
- LBW can be easily automated with robotic machinery for large volume production.
- No electrode is required.
- No tool wears because it is a non-contact process.
- The time taken for welding thick section is reduced.
- It is capable of welding in those areas which is not easily accessible.
- It has the ability to weld metals with dissimilar physical properties.
- It can be weld through air and no vacuum is required.
- X – Ray shielding is not required as it does not produce any X-Rays.
- It can be focused on small areas for welding. This is because of its narrower beam of high energy.
- Wide variety of materials can be welded by using laser beam welding.
- It produces weld of aspect ratio (i.e. depth to width ratio) of 10:1.

Disadvantages

- Initial cost is high. The equipment used in LBW has high cost.
- High maintenance cost.
- Due to rapid rate of cooling, cracks may be produced in some metals.
- High skilled labour is required to operate LBW.
- The welding thickness is limited to 19 mm.
- The energy conversion efficiency in LBW is very low. It is usually below 10 %.

Application

The laser beam welding is dominant in automotive industry. It is used in the area where large volume production is required.

Hybrid Laser Welding

Hybrid laser welding is a process, where laser welding and some other welding process, such as MAG welding, are combined to a single process. Laser beams and electric arcs are quite different welding heat sources, but both work in a gaseous shielding atmosphere at ambient pressure that makes it possible to combine these heat sources to a unique welding technique referred to as hybrid laser-arc welding. Hybrid laser-arc processes, especially hybrid laser-MAG welding, have been increasingly applied to welding applications during recent years. In hybrid laser-arc welding process most of the process mechanisms of autogenous laser welding and arc welding can be observed. Furthermore some synergistic effects originating from the combination of two different processes are present.

Hybrid laser-arc welding is a combination of laser welding and arc welding. The processes create a common weld pool which moves along the weld pass. In some applications, the distance between laser beam and arc is so long, that the processes are actually creating separate weld pools. This process is sometimes referred as tandem laser-arc welding rather than hybrid laser-arc welding. Hybrid process is not just a simple combination of laser beam and arc, but there are complicated physical interactions between the two heat sources.

Development of Laser-arc Hybrid Welding

In recent years, laser welding has begun to be used for assembly welding of automotive bodies and parts, although it has not yet to be used widely. One restraint is that in laser lap welding, the gap between the lapped sheets must be controlled very tightly. If the gap is wide, burn-through occurs, and, if the gap is excessive, the two sheets cannot be welded together. For this reason, the gap is generally restricted to 0.1 mm or less for laser welding. Laser lap welding of zinc-coated sheets poses additional problems. Laser lap welding of zinc-coated sheets is performed without any gap, so the zinc that is evaporated between the sheets tends to blow off weld metal, or the zinc vapor tends to remain in the weld metal and form blowholes.

Laser-arc hybrid welding was developed to solve these problems. This method combines YAG laser welding and arc welding, allows a larger gap between lapped sheets than in laser welding, and produces fewer blowholes, even in lap welding of zinc-coated sheets. Therefore, stringent gap control is not necessary for lap welds, and the industrial application of this method is easy. In addition, the welding speed can be equivalent to that of laser welding, so the high efficiency of laser welding can be utilized.

Configuration of Laser-arc Hybrid Welding

Fig.1 shows the system configuration for laser-arc hybrid welding. A YAG laser is used for laser welding, and an arc-welding electrode is positioned behind the YAG laser radiation point. The aim position of the arc is about 1to 3 mm behind the laser radiation point. A YAG laser is used because the plasma does not absorb much of the laser energy. Most of the energy reaches the sheets, so that the YAG laser energy is efficiently utilized for welding. In contrast, energy from a carbon-dioxide laser is strongly absorbed by the arc plasma, so that a sufficient distance must separate the arc and the laser radiation point. Therefore, the combined effect of the laser and arc is not possible with a carbon-dioxide laser.

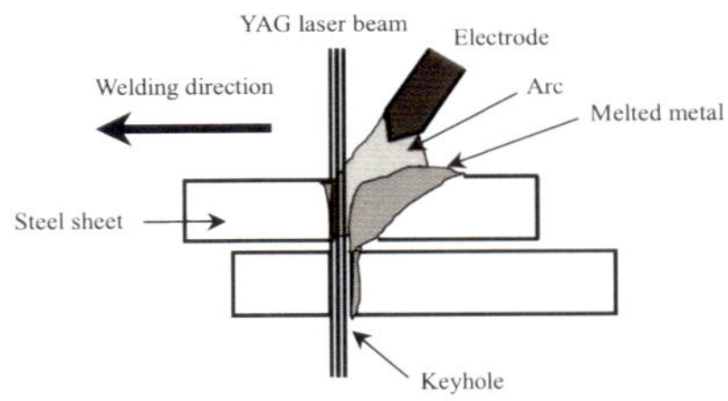

Fig: Configuration of laser beam and arc-welding electrode in laser-arc hybrid welding

In hybrid laser-arc welding processes the laser beam as a high-energy density heat source commonly serves as the primary heat source enabling deep penetration mode welding while the arc as the secondary heat source improves process stability, reliability and efficiency as well as the quality of the weld. As a secondary heat source, either process with consumable electrodes, i.e. MAG process, or process with non-consumable electrodes, i.e. TIG process, can be chosen. If any addition of filler material is required, a processes with consumable electrode should be applied, otherwise arcs with non-consumable electrodes are preferred. Depending on the energy ratio of the two heat sources, the characteristics of the hybrid process may be either more arc-like or more laser-like. The weld penetration increases with an increase in laser energy, whereas the weld width increases with the arc energy. For different applications, there exists an optimum ratio between the powers of the laser beam and arc.

Advantages

Hybrid welding is already being used, or under consideration, by the following industries:

- Road transport – the high welding speed of the hybrid process is attractive to the high production volume environments found in the automotive industry, especially given the greater part fit-up tolerance than that of autogenous laser welding.
- Shipbuilding – the lower heat input and distortion introduced when hybrid welding, when compared with MAG welding or SAW, reduces the costs associated with distortion correction and rework. By contrast, conventional arc welding methods and their associated re-work have been estimated to constitute up to 20-30% of overall manufacturing costs.
- Rail transport – as in shipbuilding, the low distortion that can result from hybrid railcar seam welding, compared with conventional arc welding processes, is of interest as a means of reducing fabrication costs, as well as the higher welding speeds reducing overall welding time.

- Oil and gas – the hybrid welding of pipes has been well demonstrated, and with continuing developments in laser sources and pipe steels, continues to be of interest as a future means of increasing overall joint completion rates, depending on steel grades used and operating environment requirements.

The chief benefits of hybrid laser-arc welding can be summarized as:

- Improved tolerance to joint fit-up: for example, hybrid welding can extend the tolerance to joint gap by a factor of at least 2-3 times over laser welding, or greater, if real-time adaptive control of hybrid welding parameters is carried out.
- Improved weld quality: hot cracking (e.g. in some higher strength Al alloys) can be avoided, and internal porosity content reduced, with respect to laser welds.
- Increases in single pass penetration depth: this is controlled principally by the choice of laser and welding parameters used, but single pass penetrations >6-12mm can be achieved using higher power (≥5kW) lasers.
- Increases in welding speed: this is also dependent on the laser used and materials being welded, but speeds of >5m/min can be possible in thinner materials.
- The increases in penetration depth and/or welding speed, when compared with arc welding, are particularly significant: the net heat input can be reduced, resulting in lower distortion, rendering hybrid welding of particular interest for the making of long seam welds between plates or between sections, the welding on of attachments etc.

Laser Shock Peening

LSP is a mechanical (cold working) process where pulses hit the surface with high power intensity and shock waves are generated. These waves plastically deform the surface and compressive stresses are extended into the subsurface. These dynamic compressive stresses are highest on the surface and decrease with depth into the material.

Generation of Laser Shock Waves

The basic steps for laser shock wave generation can be described as the following sequence; (a) the target surface is covered with an absorbent (sacrificial) coating. This layer vaporizes, forming plasma on the surface with short duration pulse pressure. Absorbent layer prevents melting and laser ablation while maintaining high surface quality and without this layer, the energy conversion from pressure to shock cannot be made effectively. This layer can be aluminum, copper, lead, vinyl tape, zinc or black paint. Hong *et al.* showed experimentally that black paint layer has the best absorption ability to laser; where almost 100% laser energy is absorbed by black paint as compared to 80% absorption by Al. This means that when a target is irradiated, almost 100% of the laser intensity is used to generate the plasma; (b) transparent overlay (tamping layer) is applied to prevent the plasma from expanding away off the surface thereby increasing the intensity of the shock wave. These overlays, also known as the confining medium, can be water, quartz or glass. Other confining media are K9 glass, Pb glass, perspex or silicon rubber. The choice of confining medium solely depends on the substrate material, density and acoustic velocity which gives the so-called “acoustic impedance effect”. Finally, plastic deformation occurs when the shock wave stress exceeds the strength (dynamic yield strength, σdyny) of the metal. This LSP process is schematically shown in figure below. The plastic deformation results in high dislocation multiplication and movements which affect the microstructure and properties of the material. Peyre *et al.* showed that increasing laser shock pressure increases the depth of plastic deformation. The description of how shock waves are generated is simple compared to the complexity of the process parameters and optimization.

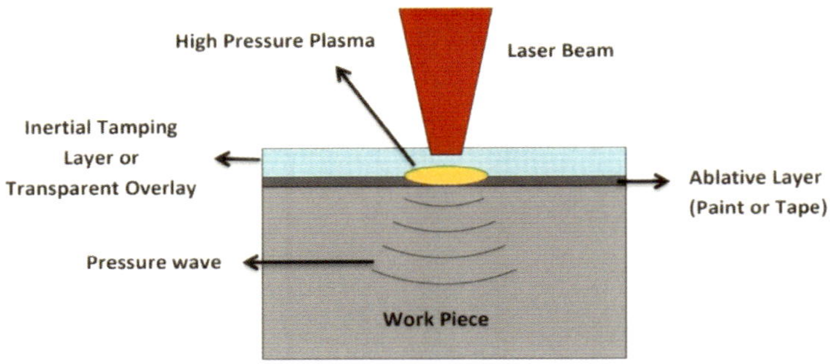

Figure: Laser shock peening (LSP) process

Process Parameters for Laser Shock Peening

Effective shock peening process depends on the material (target), laser and beam parameters as well as the absorbent and transparent overlays properties. Typical requirements are: Q-switched laser system based on neodymium such as Nd-doped glass (Nd:Glass), yttrium aluminum garnet (Nd:YAG), yttrium ortho vanadate (Nd:YVO_4) or ytterbium-doped YAG (Yb:YAG) lasers. Different laser wavelengths of 1054 nm (infra-red), 532 nm (green), and 355 nm (ultraviolet) are used. Additionally, time duration of shock wave pressure (laser pulse length) ranges from 10 to 100 ns with energy of 1–100 J per shot, and spot (diameter) size of 1–6 mm. For the laser spot geometry; circular, elliptical, rectangular and square geometries have been studied with square showing better uniformity in intensity and good overlapping rate. A frequency of less than 1 kHz and power density (I) in the range of 0.1–10^6 GW/cm^2 are also required. Comparing with UIP, a higher frequency of 15–100 kHz is employed. Furthermore, the most important LSP process parameters and equations have been summarized. Power density is expressed as a function of frequency, pulse time, power and spot area as given in first equation. Other parameters are the laser fluence as given in second equation, the load or pressure (P), which must exceed the dynamic yield strength for plastic deformation given by different authors in third to ninth equations. Tenth equation shows the reduced shock impedance (Z) for the target material (Z_1) and confining medium (Z_2) which is related to the density and speed of sound in the material. Table below lists typical values of acoustic shock impedance for various confining media and target materials. Furthermore, Eleventh equation shows the Hugoniot Elastic Limit (*HEL*), which is the ultimate stress a material can withstand under compression without internal rearrangement. The process optimization could be more complicated especially when many parameters are involved. For instance, shock pressure representation was inconsistent as found in the literature. Third to ninth equations indicate how the constants were varied based on the experimental observations. The variation of constants were attributed to the type of coating, confining medium and target material properties. This suggests that selecting the most effective processing parameters is paramount. The parameter selection can be achieved by extensive experimental work and simulations. Attempt has been made by *Society of Automotive Engineers* (SAE) to issue recommended parameters for laser peening of metals. Most importantly, successful laser peening experiments and simulations have been conducted using different processing parameters as summarized in table below.

Table: Typical values of Z_1 and Z_2 as reported by different authors.

Target material	$Z_1 \times 10^6$ (g/cm²s)	Confining medium	$Z_2 \times 10^6$ (g/cm²s)
Ti-6Al-4V	2.75	Water	0.17
AA7050-T7451	1.50	Perpex	0.32
SS304	3.61	Silicon rubber	0.47
Mg-Ca	0.88	K9 Glass	1.14, 1.5
AISI 4140	3.96	Quartz Glass	1.31
Cu	0.16	Pb Glass	1.54
SS321	4.00	BK Glass	1.44

Table: Typical LSP processing parameters from the literature used for different materials with concluding remarks.

Material	Power density (GW/cm²)	Pulse duration (ns)	Absorbent coating	Transparent overlay	Remarks
Ti-6Al-4V	5	10	Al Foil	Water	The fatigue life was improved by 22.2% and 41.7% for one and two successive laser shocks respectively as compared to as-received sample.
Ti-17	3, 6, 9	9, 27	Al	Water	It was found that all LSP parameters had influence on the residual stress profile. LSP had no influence on roughness, little effect on work hardening and large effect on hardness due to the compressive stresses induced.

AA2195	5	18	Al	Water	The fatigue crack growth rate of stir welded joint was reduced using LSP as compared to SP. Furthermore, the reduced crack growth rate was comparable with unwelded material.
Al2024 T3	5	18	Black paint	Water	LSP specimen showed resistance to fatigue crack growth for various notch geometries as compared to unpeened specimen.
4140 Steel	3	5	Al	BK7 Glass	WLSP and DSA showed improved fatigue life due to enhanced cyclic stability of compressive RS. The stability and reliability were attributed to the enhanced dislocation pinning effect associated with the WLSP and DSA process.
316 L SS	2.5	10	No coating	Water	The potentiodynamic polarization investigation showed improvement in corrosion potential (Ecorr) and corrosion current density (Icorr) with increase in laser pulse density. LPwC specimen showed 30%–40% increase in microhardness than unpeened specimen.
2204 Duplex SS	900, 1600, 2500 pul/cm^2	8	No	Water	Greater RS were observed with higher pulse density. LSP showed improved crack growth rate resistance and fracture toughness as compared to untreated specimen.
Cu 15 μm foil	6	10	Black paint	Water	The microscale laser dynamic forming (μLDF) showed that, materials are strengthened due to refined grains and large dislocation density which were dominant in the microstructure.
Alloy 22	10	25	Al	Water	Slitting method was used to establish a relationship between LSP parameters and RS on flat coupons. While contour method measured the spatial RS on critical geometrical features.
Inconel 600	10 J/cm^2	12	-	Water Jacket	Tilted column microstructure was observed due to overlapping of laser spots during scanning. Spherical nanoparticles with 60 nm diameter were seen due to laser ablation.
Fe-3%Si	0.1–1	25–200	Pb	Quartz	Laser shocking showed that quartz plus lead overlay gave the most deformation for a given power density. The deformation occurred by slip and twining mechanism. Most of the samples showed surface melting for 200 ns pulses.

Mg-Ca	78	5 to 7	Black paint	Water	Sequential peening showed an increase in the tensile pile up region up to 50% which was considered applicable for orthopaedic applications.
AZ31B Mg	5	23	Al7075	Water	Sub-grain sizes of 5.8 μm were obtained after four laser impacts. The RS near the surface increased with number of impacts. LSP showed retardation to SCC crack initiation and propagation.
Brass H62	1000, 2000, 3000 Pulses/ cm^2	10	-	Water	LSP showed that, the higher the pulse density the higher the microhardness, roughness and wear resistance. No observable features in the microstructure after LSP.
CMSX, AM1, Astroloy	12–150, 90, 25	25, 33, 40	Black paint	Water, Glass	The deformation microstructure and macroscopic mechanical phenomenon studies were envisioned to provide optimum process parameters applied to fatigue and fretting fatigue specimens.

Intensity :

$$I\,(\frac{GW}{cm^2}cm^2) = \frac{P_{avg}}{f\ (pt)a}$$

Laser Fluence:

$$Fluence\ (\frac{J}{cm^2}) = \frac{Laser\ pulse\ energy\ (J)}{focal\ spot\ area\ (cm^2)}$$

Pressure:

$$P = A\,(\frac{\alpha}{2\alpha+3})^{1/2} Z^{1/2} I^{1/2}$$

$$P(GPa) = 0.01(\frac{\alpha}{2\alpha+3})^{1/2} Z^{1/2} I^{1/2}$$

$$P(GPa)\ = 1.02 I^{1/2}$$

$$P(Pa)\ = 0.25(IMA)^{1/2}$$

$$P = 5.47(\frac{\alpha}{2\alpha+3})^{1/2} I^{1/2}$$

$$P\ (kbar)\ = BI^{1/2}$$

$$P = (AZI)^{1/2}$$

Reduced Impedance:

$$\frac{2}{Z} = \frac{1}{Z1} + \frac{1}{Z2}$$

Hugoniot Elastic Limit (*HEL*):

$$HEL = \frac{1-\theta}{1-2\theta}\sigma_y^{dyn}$$

where, I is the laser power density in GW/cm^2, P_{avg} is the average power output in W, f is the laser frequency in Hz, pt is the pulse time in ns, a is the laser spot area in cm^2, B is 21 or 10.1 for glass- or water-confined modes, respectively, A is absorption coefficient for surface coating, M is transmission coefficient for transparent overlay, α is the ratio of thermal to internal energy, Z is the reduced shock impedance of the target material and confinement medium, θ is Poisson's ratio and σdyny is dynamic yield strength at high strains.

Typical processing parameters presented in Table above are power density, pulse duration, absorbent coating and transparent overlay. However, despite the fact that the same material is laser peened, the processing parameters varied for different authors. This indicates that the choice of processing parameters is solely confined to laser type, substrate condition and target application. The characterization of LSP treated surface will be instrumental in understanding the LSP process.

Laser Peened Materials

Several materials have been peened successfully proving the effectiveness of LSP and the associated residual stresses induced. Typical research studies conducted are influence of LSP process on residual stress prolife, microstructure, mechanical properties, fatigue life, fretting fatigue life, corrosion and stress corrosion cracking behavior. Recent LSP studies have shown that laser peening technique can be used for shaping and forming of metals and inspection of coatings. The materials laser peened include titanium alloys, aluminum alloys, different steel grades, copper alloys, zinc, nickel based alloys, super alloys, brass, magnesium alloys, bulk metallic glass and other materials. Comparatively, similar materials have been treated by the UIP technique such as soft steel, 304SS, Type 35 carbon steel, 316L SS, Inconel 718, naturally exfoliated Al7075-T6511, AlSi11Mg, Al2024-T351, Al matrix composite reinforced with AlCuFe or Ti fine powders, Ti alloy, welded joints of steel grade 345W, steel grade A36, 16Mn steel and 304SS. Despite treating similar materials only a few researchers compared LSP and UIP processes.

Characterization of Laser Shock Peened Materials

Due to the high strain rates of $10^6 \cdot s^{-1}$ involved in the laser shock peening process, it is usually accompanied with changes in microstructure and mechanical properties (hardening and strengthening). Despite the high strain rates during LSP, the total induced strain is much less than in the case of shot peening. Furthermore, these microstructural changes are characterized and investigated by means of TEM and SEM for the surface and sub-surface morphologies and features. For the TEM technique, typical information revealed are dislocation structure, dislocation entanglement, dislocation band and precipitate structure. SEM technique on the other hand, provides surface and subsurface features such as surface micro cracks, macro cracks, fretting cracks, mild indentations, melting due to ablation and other surface irregularities from the LSP process. The measured compressive residual stress induced during the LSP is an important indicator of the effectiveness of the LSP process. Since residual stresses cannot be measured directly due to its extrinsic nature, strain in the crystal lattice which is an intrinsic property is usually measured. The residual stress are calculated from the strain while assuming a linear elastic distortion of the crystal lattice. Several methods have been used to quantify the residual stress profile in the material. Among these methods are X-ray diffraction $\sin^2\psi$, hole drilling, slitting, neutron diffraction and contour methods. X-ray diffraction $\sin^2\psi$ method is the most commonly used method to quantify the residual stresses. In this method, the strains are measured at particular diffraction angles and lattice planes where Bragg's law is satisfied. Material constants such as Young's modulus, Poison's ratio as well as different tilt angles "ψ" are further employed to quantify the stress. For in-depth or subsurface residual stress profiles, measurements after material removal by electro polishing is used extensively. The step-by-step material removal is in the order of microns and more details of this procedure have been reported in. For the hole drilling method, a hole is drilled and variations of stresses are measured as a function of depth by a strain gage rosette. Surface roughness measurement after LSP has been reported as the arithmetic average (Ra) using surface profile meters such as Talysurf contacting surface and electronic contact (stylus). The roughness can be areas of stress concentration and crack initiation sites which have adverse effect on fatigue life of components. Micro hardness and nano hardness across the irradiated surface are also measured and reported in the literature. Increase in measured hardness has been attributed to plastic deformation resulting in dislocation multiplication and movement. For in-depth or subsurface hardness measurements, nano hardness is a more reliable approach as compared to micro hardness. This is because of the compressed layer thickness which might pose measurement difficulty when measuring micro hardness. However, these characterization techniques give researchers an insight regarding the changes and modifications due to LSP processing. The general improvements can be further quantified after a series of mechanical testing depending on the application, such as test for fatigue resistance, fretting fatigue resistance, stress corrosion cracking, and water erosion resistance.

Advantages of Laser Shock Peening

One advantage is that in laser peening, the depth of compression exceeding shot peening produced with low cold work. The subsurface compressive residual stress distribution tends to be maximum at the surface, diminishing nearly linearly with depth. The low cold work provides thermal and mechanical stability. an advantage in hot section engine applications, or where there may be momentary overload due to impact, etc.

Disadvantages to Laser Shock Peening

Disadvantages associated with laser peening include:

- High capital cost of the laser peening equipment, which may include a clean room environment and million dollar plus installations.
- Laser shock peening requires repeated coating with tape to produce the depth and magnitude of compression achievable in a single pass.
- Because the explosive laser shocks remove the tape from adjacent treatment sites, two coating cycles are usually required to achieve one layer of treatment. Multiple layers are generally required to achieve the desired depth of compression.
- Process quality control has slowed the application of laser shock peening due to the difficulty of controlling the processing variables involved. Variation in the thickness and even turbulence of the water layer tamp, as well as debris in the air and on the mirrors affect the compression achieved, even when the laser power is precisely controlled. The laser shock peening process cannot provide the closed-loop control available with LPB. Internal cracking caused by the superposition of shock waves from opposing faces of blades, or echoing waves from the opposite wall of thicker sections has been found to limit the application. The cost of laser shock peening is the highest of any of the surface enhancement methods, typically 10 to 100 times shot peening.

Laser Engraving

Laser engraving is the process of using a laser to etch or mark the surface of a work piece. Compared to other engraving methods, laser engraving is often considered to have many advantages, though it also has some disadvantages. For instance, it can be extremely precise, but equipment costs are often prohibitive. This technology can be useful for many applications and can utilize many materials. Engraving methods vary, depending on the type of work piece and the material, which means that different engraving equipment is sometimes required.

Uses for laser engraving are usually based on the production of a specific design. The type of design often dictates the use. Professional engraving can include industrial or commercial production of engraved products. These can include duplicate images or customized images at nearly any image quality, though high-quality engraving is often more costly than low-quality engraving. Partly because of the wide range of materials that can be used, engraving also can be used for artistic work.

Stone, glass, metal, plastic and natural materials can all benefit from laser engraving. In fact, for some of these materials — such as stone and metal — engraving can be the primary form of marking for any purpose. Harder materials such as stone and glass are often difficult to engrave using other forms of engraving. For instance, laser crystal engraving relies heavily on the use of lasers to create images inside of hard, crystalline material. The variety of materials used results in there being many different types of engraving.

Industrial engraving, for instance, might make use of large machines to mass produce a single laser-engraved product or design. Professional engraving, on the other hand, might rely on small machines to customize products based on consumer needs. Despite these differences, the hardware used in laser engraving is often similar.

The hardware used in laser engraving is often constructed in three major parts: the laser, the working space and the interface. Lasers can be constructed in many ways, and some machines might have complex lasers while others rely on simple lasers. Working space often depends on the type of laser being used. For instance, simple engraving equipment might have a flat table for the work piece, whereas complex engraving equipment might hold and move the work piece on multiple axes. The interface of these systems is usually a computer into which the specific designs are put and through which the engraving commands are carried out.

Working of Laser Engraving

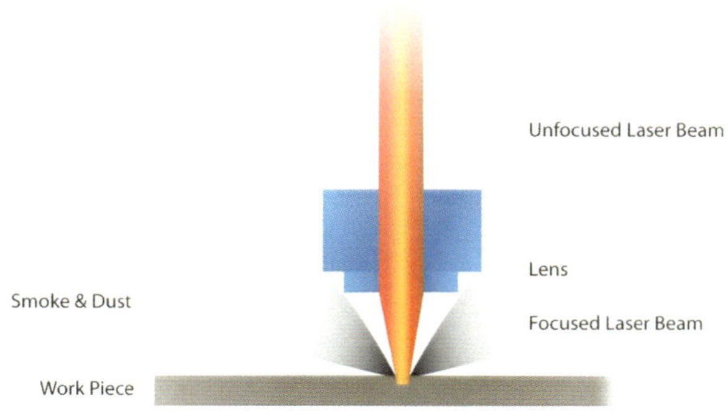

The high heat of the laser beam vaporizes the material thereby cutting into the part's surface and physically removing material. This process leaves a cavity in the surfaces that is not only visible with a high contrast but also noticeable by touch. There are noteworthy differences in the depth of this cavity that varies between 0.02″ in metals to 0.125″ in harder materials.

Usually, the engraved areas become black. Multi-layer materials offer an alternative, as they enable to engrave other colors than black. By removing the top layers the lower layers become visible.

Another variant of laser engraving is relief engraving. The laser creates a deep engraving with distinct heights. This variant is beyond any doubt the most impressive, since the overall effect is that of a wooden sculpture.

Laser Engraving on Different Material

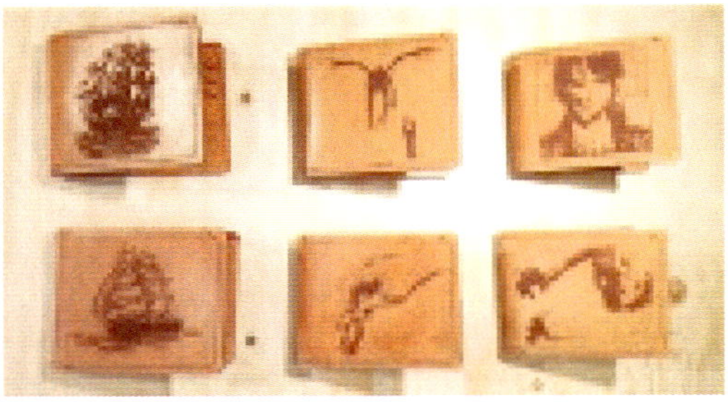

It is possible to laser engrave finished leather, producing a look similar to hot-branding. Avoid raster marking large surfaces, since the leather will turn powdery. Therefore, vector marking is recommended.

Since leather is a natural material, it may deform during laser engraving if too much power is applied.

Because of the many varieties of leather it is difficult to come up with a general rule for laser engraving it. Hence, there is no way around determining the optimal laser engraving settings for each kind separately. To this end test your machine with a gray scale matrix

Laser Engraving Paper/Cardboard

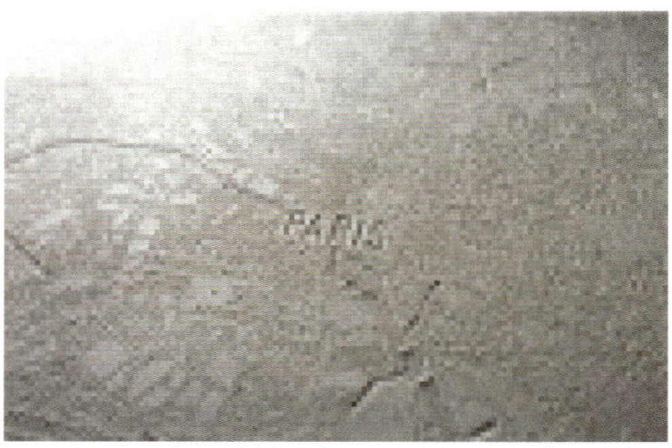

Because of the many varieties of paper and cardboard it is advisable to determine the optimal laser engraving settings for each kind separately. To this end test your machine with a gray scale matrix.

Laser Engraving Wood

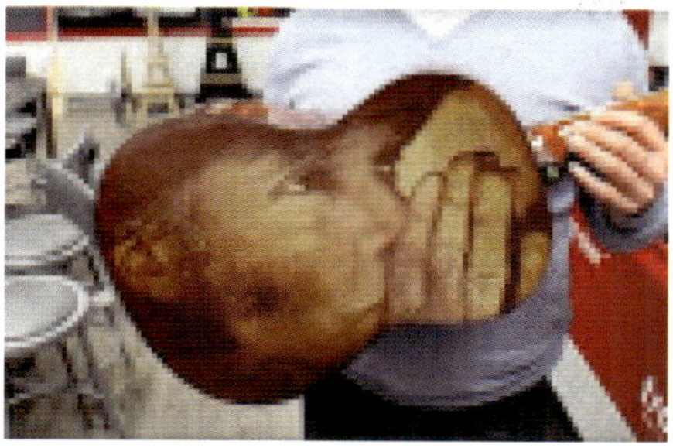

Besides natural woods there is also plywood, which has several advantages. This laser engraving material consists of several sheets of wood that are glued together. As a result, plywood boasts a raised mechanical resistance, water resistance and lightness. All these properties have made plywood a staple for builders, designers, artists, and craftsmen.

In terms of laser engraving, plywood has an additional advantage over natural wood. When cut into sheets, the latter reveals grains in alternating shades that stem from different kinds of growth during the seasons. These differences also translate into different laser engraved surfaces. Usually, the lighter areas will come out lighter and the dark areas will be darker.

As a rule of thumb, unless the desired look of your design is kind of zebra pattern, you will be better off using plywood that has a uniform surface.

In addition, you should always apply higher laser power when engraving hard wood. Only then will you smooth results.

To determine the optimal laser engraving settings, test your machine with a grayscale matrix.

Laser Engraving Glass

The effects the laser has onto the glass surface can be manipulated using a number of practical methods.

Apply a moist paper towel to achieve a smoother surface finish and a white engraving result. Take care that there are no air bubbles or overlaps, because this would distort the effect. Once the engraving is finished the paper towel can be wiped off easily.

An alternative to paper towels is application tape. Like moist paper towels, application tape is an excellent measure against surface roughening. The same caveat concerning air bubbles and overlaps apply. However, the final engraving color is not a bright white, but looks grey. Once the engraving is finished the tape residue can be removed off easily.

But what to do when your laser engraver has not enough power to engrave glass? Even then you can achieve designs with high-contrasts! Simply mask the glass surface with tape. Then you engrave the design into the tape. Next, cover the revealed area of the tape with a thin layer of paint using a brush. After the paint has dried, peel the tape and remove the tape residue carefully.

Laser Engraving Stone

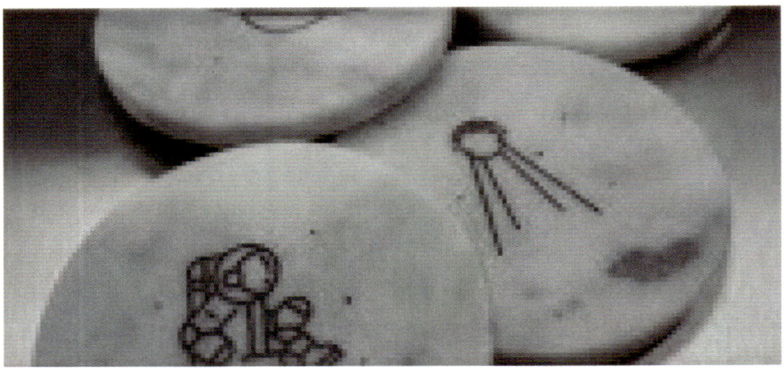

Experience shows that it can be piece of work to laser engrave stone. You will get the best results with polished stones. To avoid distortions of the laser engraving motif, try to use stones that have a flat surface. You can align natural stones with the laser by using modeling clay as a support. When you set the laser's focus, apply an average value to achieve a uniform result.

If your results are poor, consider changing the Z-offset value. This simple trick can significantly improve the result of your engraving. Many professional laser engravers report that hard stones are easier to process if you choose a negative Z-offset value of 1mm or even more. The physics behind this workaround are simple. The closer the surface to be engraved is to the laser, the more energy enters.

But what to do when your laser engraver has not enough power to engrave stone? Or, when working with extremely hard stones that yield no satisfying results? Even then you can achieve designs with high-contrasts! Simply mask the glass surface with tape. Then you engrave the design into the tape. Next, cover the revealed area of the tape with a thin layer of paint using a brush. After the paint has dried, peel the tape and remove the tape residue carefully from the stone's surface.

Laser Ablation

Laser ablation is the process by which layers are removed from solid metals and industrial compounds using a laser beam for ultimate precision. The beam will irradiate the surface, meaning that it has been exposed to radiation.

Laser Ablation Process

Laser ablation works by focusing a laser onto a substrate to a remove material that is on its surface. The amount that is removed depends on the intensity, pulse length, and wavelength of the laser, as well as the material itself. The area absorbs the laser that is being directed on it, thereby breaking down the chemical bonds within the area.

With a low level of laser flux, the material being focused upon is absorbed by the laser's energy and then changes to a gaseous state. With a higher level of laser flux, the material that is being focused upon will usually be converted to plasma. It is possible to undergo the laser ablation process with both a pulsed fiber laser and a continuous wave laser, although the former is the more common method due to the high level of laser intensity.

Laser ablation has many benefits over more traditional methods of processes, such as with thin film removal, whereby alternative solutions have to undergo a multi-step process which is costly, time-consuming and inflexible, as well as having risks for the environment. Laser ablation is a much more efficient, reliable and cost-effective method.

Uses of Laser Ablation

Laser ablation has many different uses, particularly as the whole process is undergone with minimal excess heat being transferred to the surrounding area of the material being used. This means that the process will have little to no effect on the parts of the material that you don't want it to, and so laser ablation is great for a wide range of materials, including plastic, metal, ceramic, and even biological tissue. With the advanced nature of fiber lasers, you can focus on a very small, specific spot to produce your desired results.

Laser ablation can also be used to determine the presence and concentration levels of a particular chemical or material on a surface. This is achieved by generating bright plasma on the surface, and then analysing this plasma to see what is present there. This is a much more environmentally-friendly process for determining chemical analysis, as opposed to more traditional methods such as using toxic acid solutions.

Finally, laser ablation can be used to deposit film on a surface, so as to produce things like Nanomaterials or to fabricate superconducting materials. This can also be used for the micro-structuring of an object, whereby the structuring of a larger object can be seen.

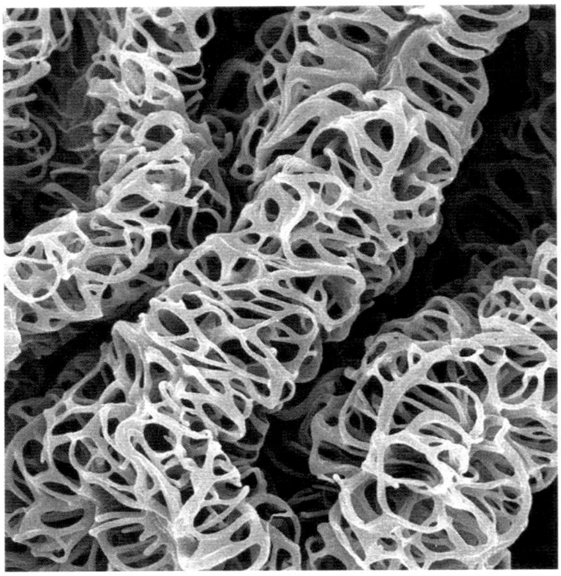

Fig: Laser ablation is also useful for viewing the micro-structuring of an object

- Factors Need to be Considered for Laser ablation

The success and efficiency of the laser ablation process depend on a number of factors. These are:

- Wavelength

The wavelength needs to be selected carefully with a minimum absorption depth. This is so that there is a high energy deposition within a small volume, resulting in a quick and efficient laser ablation.

- Pulse duration

It is better to use shorter pulse durations so as to perform at maximum peak power and to minimise the thermal damage that is inflicted on the surrounding area, although this is already greatly reduced due to the precise nature of the laser ablation process.

- Pulse repetition rate

The pulse repetition rate needs to be high enough so that the heat caused by laser ablation is maintained and is not left any time to cool. This will result in a more successful ablation as energy waste will be reduced.

- Beam quality

Your laser beam needs to be of sufficient quality to perform the laser ablation process successfully. The quality of the beam will be determined by its brightness, its focus ability, and its homogeneity. The beam size will also need to be controlled so that you don't ablate too large an area.

- Application Areas for Laser Ablation

Due to the precise, accurate and environmentally-friendly nature of laser ablation, it is used in many different application areas and industries. Its ability to remove thin film has particular benefits for the electronics and semiconductor industries, and it has also found uses for other industries such as medical, automotive and ship-building. It has understandably become the preferred solution for many manufacturing processes.

Laser Bonding Technology

Laser bonding is an established technology that uses lasers to bond additive marking material to a wide variety of different substrates and is commonly used with CO_2, Nd: YAG and fiber lasers.

The technology was first invented in the mid 1990s by Paul W. Harrison, the founder of The Mark. After selling his interest in The Mark, he went on to found Laser Bonding Technology and continues to be involved with laser marking hardware and the further development of new and unique marking materials. These new products produce permanent marks on most metal, glass, ceramic and stone surfaces for use in a wide range of industrial and artistic applications, ranging from automotive, aerospace and medical to the awards & personalization industries. It differs from the more widely known techniques of laser engraving and laser ablation in that it is an additive process, adding material to the substrate surface to form a permanently bonded mark or decoration instead of removing material as is done in those other laser processes.

For metals, parts can be permanently marked with high contrast and high resolution for identification, logos, bar-coding and serialization purposes, without damage to the substrate. With glass and ceramics, complex surfaces can be decorated or marked; however, the traditional kiln firing process is replaced by a laser, with its permanently bonded decorations and markings fired on in seconds instead of minutes or hours.

References

- Electrical-discharge-machining-EDM-424360: britannica.com, Retrieved 11 May 2018
- Laser-beam-welding: theweldingmaster.com, Retrieved 21 March 2018
- Hybrid-laser-welding, laser-processing, technologies: ionix.fi, Retrieved 29 April 2018
- Laser-shocking: lambdatechs.com, Retrieved 19 June 2018
- What-is-laser-engraving: wisegeek.com, Retrieved 10 July 2018

8

Laser Cutting Technology

Laser cutting is the technology of cutting materials using lasers for industrial manufacturing. The topics elaborated in this chapter which include laser cutting machine, laser vaporization cutting, flame cutting, laser fusion cutting, laser oxygen cutting, etc. help to provide an extensive understanding of laser cutting technology.

Laser Cutting

Laser cutting is a way to cut precise patterns in metal, plastic, wood, and practically every other material that man works with. It allows a level of accuracy and complexity impossible with conventional machining tools. Laser cutting works by exciting a gaseous medium, commonly carbon dioxide, causing it to amplify light reflected back and forth multiple times within the laser chamber. The light emerges from an aperture and is focused by a lens onto a specific point.

A typical process laser has a beam about a 1/5 of a millimeter in width, focusing 1000 to 2000 watts of energy. This is enough to melt most common materials. Because lasers become less focused and lose energy as they penetrate through a material, there is a limit of about 20 mm for the deepness of the cut. Laser cutting machines are integrated into a larger CAD/CAM (computer-aided-design, computer-aided-manufacturing) system that takes a design file and implements it on a work piece. These machines represent a stepping stone in the continuing trend away from hands-on manufacturing, putting human workers in a more removed, creative design role.

Because a laser is made up of photons, parts of its energy can be reflected away by materials such as aluminum and copper alloys. These materials are also thermal conductors, meaning they distribute incoming heat more evenly throughout their volume. For this reason, carbon alloy and stainless steel are popular work piece materials for laser cutting. They are poor at absorbing heat, so heat is concentrated into the laser's path more readily.

Typical laser cutters allow features as small as 1 mm in size. Specialized lasers often feature sizes that are even smaller, allowing the fabrication of true micro machinery. The first laser cuts were made in 1967, as part of a spin off of a military research project focused on lasers. Because the beams used in cutting are "class 4" lasers, the machines are designed to ensure that human operators are never exposed to them directly. All the cutting is done inside the machine.

Types of Laser Cutting

There are three main types of laser cutting:

- Co_2 laser cutting

Co_2 laser cutting is done by using a gas laser. The gas in question is a carbon dioxide mixture which has been electrically stimulated. Co_2 laser cutting is most often used on non-metal materials as they have a wavelength of 10.6 micro metres.

- Crystal laser cutting

Crystal laser cutting uses nd:YAG (neodymium-doped yttrium aluminium garnet) and nd:YVO (neodymium-doped yttrium ortho-vanadate) lasers. These are crystals which belong in the solid state group of lasers, and have the same wavelength as fiber lasers, around 1.064 micrometres. This means that they are useful for both metals and non-metals.

However, this type of laser machine is made up from expensive pump diodes, which can be costly when needed to be replaced; typically, after 8,000 – 15,000 hours of use. The crystal used also has a much shorter service life than that of a fiber laser.

- Fiber laser cutting

This type of laser cutting is completed using a fiber laser, the type of lasers that we have here at SPI Lasers. Like the method of crystal laser cutting above, fiber lasers belong to the solid state group too.

A beam is created using a 'seed laser', which is then amplified using glass fibers and pump diodes. They have the same wavelength as above, 1.064 micrometres. This means that the laser beam for a fiber laser is roughly 100 times greater than that of a Co_2 laser when it comes to intensity. It also means that they can work with both metal and non-metal materials.

A fiber laser is the most useful of the three laser types. Alongside the benefits listed above, they are often maintenance free, require much cheaper replacement parts, and have a much longer service life of around 25,000 usage hours.

Materials that can be Cut by Laser Machine

A laser cutting machine is great at cutting a huge variety of materials. Listed below are just some of the many materials:

- Wood
- Gemstones such as diamonds
- Titanium

- Steel
- Reflective metals
- Glass
- Plastic
- Silicon.

Whatever the material is that you need to work with and whatever the application, it's highly likely that laser cutting will more than be up for the task.

Characteristics of Laser Cutting

Laser cutting is most commonly used laser beam machining (LBM) process due to its different advantages such as narrow kerf, small HAZ, high accuracy and cut edge quality, material versatility, high flexibility, no tool wear, non-contact nature, high production rates and better material utilization. Due to excellent cut quality and narrow kerf, laser cutting process can be applied for a very precise cutting of sheet metal. The intricate and complex profiles with very narrow kerf width can be created in any type of material by the laser cutting process. The concentrated laser beam produces very little HAZ and laser cut specimens do not require further subsequent machining operations. Laser cutting can be applied to almost all categories of materials such as metals, super alloys, ceramics and composites. The laser cutting capability depends mainly on optical and thermal properties of the material rather than mechanical properties. The materials which exhibit high degree of hardness or brittleness and possess through favorable thermal and optical properties such as low reflectivity, low thermal conductivity and diffusivity are well suited for laser cutting process. As laser beam can be transferred from one place to another place with the help of optical fiber, it can be applied for cutting in inaccessible areas.

According to the different types of lasers are available in market, solid state Nd: YAG and gaseous CO_2 lasers are mostly used for cutting in industries due to their high powers and other suitable properties. The use of CO_2 lasers is not common in practice for cutting of highly reflective and thermally conductive materials such as copper, aluminum and their alloys due to its poor absorptivity. High reflectivity of these materials require high laser power when cut with CO_2 laser and there is possibility of damage to laser cavity, cavity optics or beam delivery optics due to reflected laser beams. Further, high thermal conductivity of aluminum alloys tends to produce large heat affected zone (HAZ), wider kerf and oxides on the molten materials in the cutting front. Although Nd:YAG lasers have low average power but when operate in pulsed mode, give higher peak power and beam intensity. Due to the shorter wavelength (1.06 μm), it can be better absorbed by reflective materials such as copper, aluminum and their alloys as compared to the CO_2 laser. Pulsed Nd: YAG laser cutting becomes an excellent means of cutting due to good focusing characteristics, smaller kerf width and narrow HAZ.

Most of the organic materials such as plastics, wood based products, leather and natural rubber cannot be cut by Nd: YAG laser as these materials are transparent to the Nd: YAG laser light and most of laser light passes through the material without heating it. But CO_2 laser can be used for the cutting of these materials.

Applications of Laser Cutting

The main industrial application of the lasers at the present time is in the laser cutting of different materials. The applications of laser cutting in different fields are numerous. Laser cutting is the neatest and the fastest profile cutting method. Since the setup time is very less, the jobs can be completed in only few minutes. The ease of the profile cutting has opened up several interesting novelties in design. It is used in an economical way for the production of many components. Laser cutting is widely used in different technologically advanced industries such as automobile, aerospace, nuclear, medical, chemical, civil structures and household applications. Mild steel sheets are most commonly used for civil structures and house hold applications which are easily cut by laser cutting.

Laser cutting is used for the cutting of titanium and its alloys which are most widely used for different industries such as aerospace, marine, chemical, food processing and medical due to their superior performance characteristics such as high strength and stiffness at elevated temperatures, high strength to weight ratio, high corrosion resistance, fatigue resistance, and ability to withstand moderately high temperatures without creeping.

Aluminium and its alloys cannot be cut easily by conventional cutting methods due to their thermal and electrical properties so that laser cutting is applied for the cutting of this type of materials. It is widely used in the market due to its unique properties such as their lightness, high strength-to-weight ratio and good corrosion resistance. These are most widely used in aircraft, automobile, ship building and chemical processing industries. In the coated form (alclad sheet), duralumin is used for the construction of ship frames, and when alloyed with nickel, it is used for making the pistons and piston rings of aircraft and automobile engines.

Besides the metals, laser cutting is also applied for the cutting of super alloys which are commonly used for making the different parts such as aerospace parts, gas engine components, parts for nuclear power plant, ship building due to its good mechanical properties at high temperature and corrosion resistance in hostile environments. Laser cutting process is also applied for nuclear power plant to dismantle and repair the plant with the help of high power lasers.

Apart from these applications, laser cutting is applied for the cutting of difficult-to-cut materials such as ceramics and composites which have many applications in mechanical, electrical and engineering applications due to outstanding thermo-mechanical and thermo-chemical properties. The traditional methods of ceramic tiles cutting, such as diamond saw, water jet machining and ultrasonic machining, are very expensive and time consuming. Laser cutting is applied for the cutting of ceramic tiles.

Short and ultrashort pulse lasers are used for the cutting of different parts at micro and nano levels. Short pulse Nd: YAG laser is successfully used in electronic industry to cut QFN packages (semiconductor packages which are plastic encapsulated packages with copper lead frame substrate). High quality femto-second lasers are used for the micro cutting of alumina ceramics which is used for the manufacturing of high power radio frequency electronic circuits and fabrication of multichip modules, mainly due to its high electrical resistance, high dielectric strength and high thermal stability.

In the medical field, different surgeons use lasers for the cutting nickel- titanium shape memory alloys which are used for different medical applications such as wires for minimal invasive surgery, super elastic devices for cardio-vascular systems, and spinal cord.

Bulk metallic glasses (BMGs) are disordered materials that lack the periodicity of crystalline structures. BMGs exhibit many exceptional physical, chemical and mechanical properties that differ markedly from those of crystalline metals. Thus, BMGs have attracted considerable attention as structural materials. Laser cutting with ultra short pulses is employed for the cutting of BMGs.

Laser Cutting Machine

Laser cutting machine adopts the energy released on the time when laser beam irradiate metal surface. The metal is melt by laser and sinter is blow away by gas. Because laser power is highly focused, only a very little heat effects the other part of metal plate and causes a little or no deformation. Laser can cut any complex shape precisely, which needs no further processing.

Laser source is generally CO_2 laser beam with operating power of 500～5000W. The power is even lower than that of many household electric heater, and because of lenses and reflectors, laser beams are focused in a very small bit of area. Highly focused energy heat the area quickly and makes the metal plate melted.

Laser cutting machine can cut stainless steal of thickness less than 16mm; when adding oxygen in laser beam, the cutting thickness is 8～10mm but it will generate a thin oxidation film in the cut surface. The maximum thickness is 16mm which leads to larger cutting deviation on the size of components.

Since the advent of laser, numerous laser products have been developed, such as laser printer, laser cosmetic instrument, laser marker, laser cutting machine etc. Due to its late start in China, the laser technology in China is greatly behind the developed countries. Although Chinese manufacturers can produce plenty of laser products, some key parts such as laser tube, driving motor, galvano meter and focus lens are imported products. This leads to an increase on cost thus an increase on consumer's payment.

In recent years, domestic research and production of laser products become closer to advanced overseas products with the progress of laser technology in China. Some aspects are even superior to products abroad, which has a leading role in market because of the advantages of price. Overseas products have absolute predominance in precision machining for its quality on stability and endurance.

Work Theory of Laser Cutting Machine

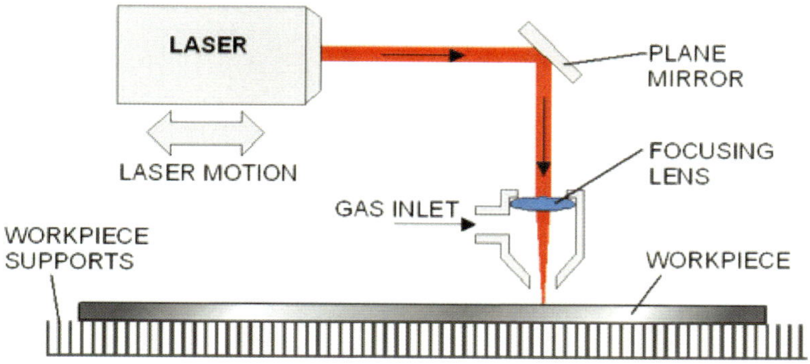

Laser tube is the core part of laser cutting machine. CO_2 laser tube.

Laser tube is composed of hard glasses, so it is fragile. It adopts layer of sleeve construction with discharge tube in the most inside layer. However, the diameter of discharge tube is thicker than laser tube, diffraction between the thickness of discharge tube and the size of flare is in direct ratio; the length of tube is in proportion to output power of discharge tube. Laser tube generates a large quantity of heat in the operation of laser cutting machine, which influences the normal work. So cold water machine is needed to cool laser tube, ensuring constant temperature for successful running.

Cutting Features of Laser Cutting Machine

- One — high efficiency

Laser cutting machine is always connected to several numerically-controlled rotary tables to achieve numerical controlled cutting. It only needs to change the NC program to adjust to components of different shapes, which can make 2D cutting as well as 3D cutting.

- Two — high speed

When cutting low carbon steel sheets of 2mm thickness, the speed of 1200W laser cutting is 600cmmin; when it is 5mm thick polypropylene resin plate, the cutting speed is 1200cmmin. The material needs no clamping fix in laser cutting process.

- Three — high quality cutting

Laser cutting features thin kerf. The two sides of kerf are parallel and the kerf is vertical to the surface. The cutting precision can reach to ±0.05mm. The cutting surface is clean and nice, with roughness of tens of microns. The cut components can even come into use directly without further machining. After laser cutting, the heat effected area is very small and material near to kerf has not been affected, making little deformation, high cutting precicion and perfect geometrical shape.

- Four — non-contact cutting

Laser cutting is non-contact cutting, which means no tool wear problem. When processing different shapes, there is no need to change tools, the only way is to alter the output parameter of laser. The whole laser cutting process features low noise, little vibration and little pollution.

- Five — various cutting material

Compared to oxyacetylene cutting and plasma cutting, laser cutting can be applied on more materials, including metal and non-metal, metal matrix and non-metallic matrix composite, leather, wood as well as fibers.

Laser Vaporization Cutting

Vaporization is the main way to remove the processed material. In the process of vaporization cutting, work piece surface is heated to vaporization temperature quickly by focused laser beams, forming high pressure steam and spraying outward at supersonic speeds. In the meantime, a hole is formed in the laser active area and laser beams reflex several times in the hole to increase the absorption of laser by material.

When high-pressure vapors spray outward, the melted materials are blown away in the kerf till the work piece is finally cut. Vaporization cutting needs very high power density, which is eighth power of ten watt above per square centimeter. It is usually applied in low flash point materials and refractory materials.

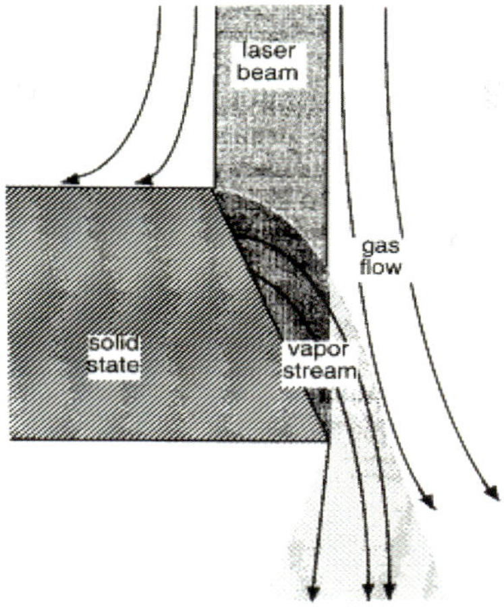

Figure: A sketch of laser vaporization cutting

Typical materials that are cut by the vaporization method are acrylic, polymers, wood, paper, leather and some ceramics. This method has a high power requirement that depends on the thermal properties of the material. High power densities are obtained by appropriate adjustment of the laser radiation and focusing. For cutting of metals, laser vaporization cutting is the method with the lowest speed among other methods; however, it is suitable for very precise, complex cut geometries in thin work pieces.

Vaporization Cutting Formulas

Recall the velocity of melt front formulas,

$$v_s = \frac{H}{\rho(CT_v + L_v)}$$

Where, H is power density absorbed per square area and the temperature at the surface from the uniform illumination formulas for vaporization point,

$$T(0,t) = \frac{2H}{k}\sqrt{\frac{\alpha t}{\pi}}$$

Thus the time for vaporization is,

$$t_v = \frac{\pi}{\alpha}\left[\frac{T_v k}{2H}\right]^2$$

If we had a 2 KW laser focused to 0.2 mm Then average power is,

$$H = \frac{2000}{\pi r^2} = 6.3\times 10^{10}\, Wm^{-2}$$

Laser Fusion Cutting

The laser fusion cutting process, also called inert gas melt shearing, is based on transformation of the material along the kerf into the molten state by heating with laser energy and the molten material blown out of the kerf by a high-pressure inert gas jet. The laser beam is the only heat source during this cutting process and the high-pressure inert gas jet is responsible for melt ejection. The inert gas jet (mainly nitrogen or argon) is also responsible for shielding the heated material from the surrounding air as well as protecting the laser optics.

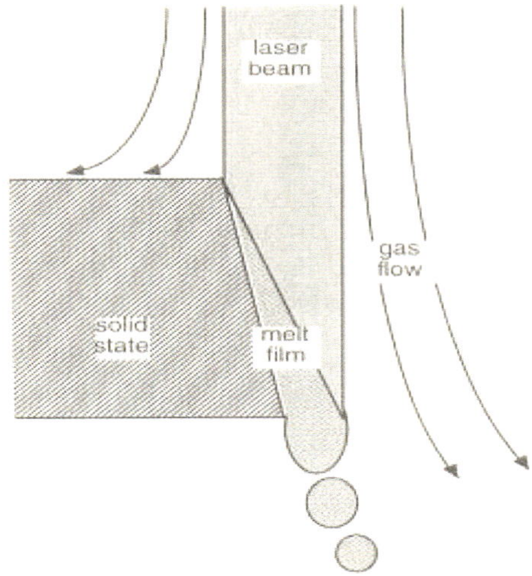

Figure: A sketch of laser fusion cutting

Laser fusion cutting is applicable to all metals especially stainless steels and other highly alloyed steels, aluminium and titanium alloys. A high quality cut edge is formed but the cutting speeds are relatively low in comparison with active gas cutting mechanisms. The advantage of this process is that the resulting cut edges are free of oxides and have the same corrosion resistance as the substrate. The cut edges may be welded without any post-cutting reparation. The main technical demand is to avoid adherent melt (dross attachment) at the bottom edges of the kerf. A high pressure (above 10 bar) is recommended to remove liquid that can adhere to the underside and solidify as dross.

Laser Oxygen Cutting

The principle of laser oxygen cutting is that the focused laser beam heats the material in an oxidizing atmosphere and ignites an exothermic oxidation reaction of the oxygen with the material. The exothermic reaction supports the laser cutting process by providing additional heat input in the cutting zone resulting into higher cutting speeds compared to laser cutting with inert gases. The laser beam is responsible for igniting and stabilizing a burning process within the kerf, and the assist gas blows out the molten material from the cut zone and protects the laser optics. Figure 16 is a schematic of laser oxygen cutting.

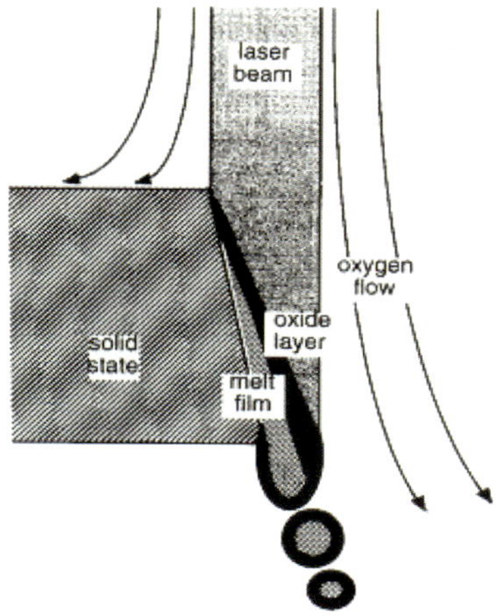

Figure: A sketch of laser oxygen cutting

Laser oxygen cutting is applicable to mild steel and low-alloyed steel. The formation of the oxide layer on the cutting front increases the absorption of the laser radiation compared to absorption of a pure metallic melt. The oxides reduce the viscosity and surface tension of melt and thereby simplify melt ejection. However, the resulting cut edges are oxidized.

Thermal Stress Cracking

Mechanism of thermal stress cutting is that laser beams heat an area of fragile material to produce evident temperature gradient. The high surface temperature makes expansion and inner lower temperature hinders expansion, forming pulling stress in the surface and radial crushing stress inside. When the two stresses exceed fracture limit strength of the workpiece, crackle appears. And then the work piece is broken along the normal direction of the crack. It is suitable for glasses and ceramics.

Stealth Dicing

In fabrication processes, wafer dicing plays an early and critical role in the quality of the final product. Two conventional methods for dicing wafers — blade dicing and laser ablation dicing – pulverize the wafer material in the cutting path (known as dicing street or kerf). In doing so, these techniques can cause problems such as debris, damage to the device and loss of precious semiconductor material.

A laser technique called stealth dicing can overcome these problems by cutting the wafer from the inside. The method involves two steps: subsurface laser-induced perforations of the wafer followed by tape expansion to separate individual chips.

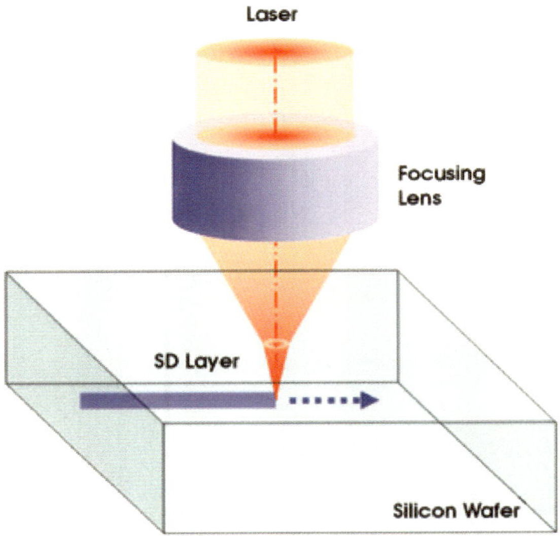

Figure: In stealth dicing, a laser beam that transmits through the wafer is focused beneath the surface, creating a row of perforations in a "stealth dicing" (SD) layer

The initial subsurface treatment uses a pulsed laser beam at a wavelength that transmits through the wafer. For pure silicon wafers, near-infrared lasers are favored over short-wavelength lasers because of the absorption of shorter wavelengths by silicon. The pulsed laser beam is focused to a point beneath the wafer's surface using an optical system. When the laser beam hits a peak power density at the focal point, the wafer absorbs the energy, and a pinpoint hole is created. The laser also causes compressive and tensile stress near the hole. As the laser beam travels the length of the wafer at a processing speed of 300 mm/s for a 120-μm-thick wafer, it perforates the inner layer of the wafer. The front and back surfaces remain pristine.

After laser processing, the wafer is still in one piece. The separation of individual chips occurs in the next step, when the tape upon which the wafer rests is expanded. The stress areas introduced by laser processing help this separation. A strong compressive stress around the hole is bounded on top and bottom by strong spots of tensile stress, and this sandwich of stress creates fissures along the laser's path – like fault lines splitting the Earth – when the tape expands. Each chip is pulled free from the others. Thus, the cutting of the wafer originates from within the wafer and not from its surface.

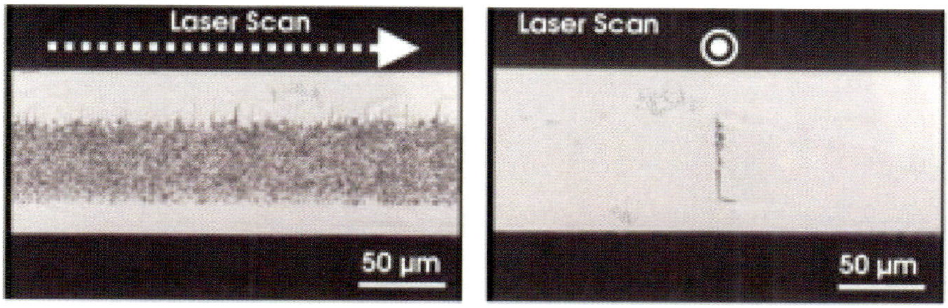

Figure: In the stealth dicing method, the laser beam creates perforations beneath the wafer's surface. The front and back surfaces are not affected

The stealth dicing technique originally was developed by Hamamatsu Photonics KK in Japan for dicing ultrathin semiconductor wafers, but it has performed well on silicon wafers of all thicknesses and on specialty wafers with die attach films, low-k materials and micro electro mechanical systems (MEMS).

The technique also may find use on other materials, including sapphire and glass. In addition, it is gentler to the environment than other methods because it is a dry process, reducing water consumption by 622 tons per year for each unit using the technology.

Dicing Silicon Wafers

Silicon wafers thicker than 100 µm traditionally are cut by diamond blades or saws, and the blade's thickness, grit size, and rotating and cutting speeds affect the cutting quality. This technique has been refined over the years, yet it still causes problems that cannot be ignored. Chipping of the wafer occurs because of the force of the sawing process. The large cut width (kerf) caused by the blade wastes the wafer and produces debris that must be cleaned up. Other issues include wearing down of the blade over time and the cost of blade replacement.

For silicon wafers less than 100 µm thick, laser ablation offers an alternative to the blade technique, which is too powerful for the delicate thin wafers. However, laser ablation has its own problems. Vaporizing the wafer with a laser along the dicing path creates molten debris and micro cracks. The debris deposited on the surface of the wafer is difficult to clean up, and the cracks result in chips with lower strength.

In contrast, stealth dicing does not generate the problems brought on by either the blade or laser ablation techniques. With stealth dicing, there is no chipping or debris, which eliminates the need for a cleanup process. It also reduces the amount of wasted wafer because of its very narrow dicing path, offering space on the wafer for more chips and higher production yield. Also, the technique does not cause heat damage because the laser cuts below the surface of the wafer; this contributes to rendering the chips more resistant to breakage.

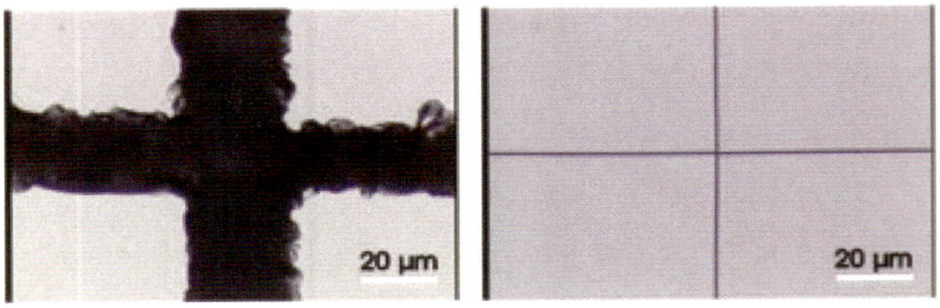

Figure: Chips are diced from a 100-μm-thick silicon wafer by the blade method at a processing speed of 100 mm/s (left) and by the stealth dicing method at a processing speed of 300 mm/s (right)

Dicing Specialty Wafers

Specialty wafers — such as those with die attach films and those made of low-k material — and MEMS devices can pose problems for blade and laser ablation techniques.

Die attach films are used as adhesives for stacking and pasting multiple thin chips in a package. For the conventional blade method, the film adds another level of dicing problems because it is cut at the same time as the wafer; this can cause warping because the wafer and film have different mechanical properties.

However, stealth dicing handles these wafers with no problem. The wafer is laser-processed first, and then the die attach film is attached to the wafer. When the tape expands, the film is separated into the exact shape of the chips because of the very high adhesion forces between the film and wafer.

Low-k materials are widely used in chips to lower the dielectric constant of interlayer insulating films to allow higher-speed operation. For dicing wafers with low-k materials (especially ones that require a completely dry process), conventional dicing processes cause chipping, debris and delamination as well as large kerf widths. Stealth dicing, on the other hand, produces a clean cut.

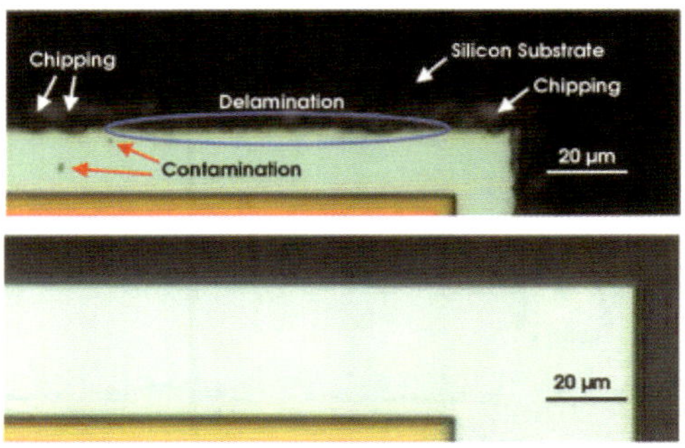

Figure: Low-k device chips are shown being cut by the laser ablation method (top) and by the stealth dicing method (bottom)

MEMS are tiny mechanical electromechanical devices built on a silicon substrate and used in a variety of applications such as biotechnology and communications. These devices are very sensitive to external impacts, which makes stealth dicing an ideal method to use because it has a low external impact. No chipping or cracks occur, and extremely sharp edges are possible.

Stealth dicing offers a dry process, high speed, high quality (no chipping or debris) and low kerf loss. It is available as a laser module that is integrated with optical systems and that is incorporated into dicing machines.

Flame Cutting

The process often called flame cutting is known by many names, such as oxy acetylene Cutting, oxy fuel gas cutting, oxygen burning, steel burning and other terms too numerous to mention. The process is now about 111-112 years old as it was patented in 1901 by Thomas Fletcher. One of the first commercial applications was what the British referred to as "an unauthorized bank entry", or a "safe cracking".

Working

The process is a thermo-chemical process requiring a source of intense heat, referred to as "Preheat", and pure oxygen. The pure oxygen requirement is of a 99.5% minimum purity, which is the minimum purity typically provided in cylinders and bulk systems. A reduction in purity of.1% will reduce cutting speed about 10%, so one can see the necessity for the purity.

While the source of the oxygen may be pure, improper connections, bad hoses, or leaks of any sort can allow impurities in the system thereby reducing the cutting speed. The high purity of the Oxygen presents a very dangerous situation and requires extreme care in the selection of equipment and the design of piping systems for its use. There are numerous documents available to assist with these safety issues. Federal Law (OSHA) presents numerous requirements for the equipment to be used, as well as the procedures to be followed.

The process can be used for cutting/severing many different materials with the basic requirement that the oxide formed must have a lower melting point than the base material to be cut. A good example of a material one cannot cut is Aluminum whose melting point is in the order of 1,200-1,300 degrees F but whose oxide is in the range of 5,000+ degrees F. Stainless steel has additives such as Chromium and Nickel added for "Oxidization Resistance" and this is the process (Oxidization) one is trying to promote, i.e. this type material is not cut without some assistance.

Thick Material can be Cut

The process can cut from sheet metal thicknesses to 100 inch material. Prior to 1958, when Plasma was introduced, all thicknesses were flame cut. Today metal thinner than 3/8-1/2 are typically only cut by plasma or laser, as thin materials are very challenging to flame cut. Thin material requires a very intense preheat as the cutting speed required for a good quality cut is very high, therefore a high temperature high velocity preheat is desired.

The process for all thicknesses is the same and that is the material must be "preheated" to a temperature of 1,600-1,800 degrees F, then the pure Oxygen is discharged into the preheated area and the steel is then oxidized or burned, hence the term "burning". (The term "burner" is often used for the operator.) The quality of the final cut surface can be quite excellent with a sharp top edge, square/flat cut surface, and a sharp slag- free lower edge.

Parameters to Control Flame Cutting

Cut quality is dependent on a number of parameters, the primary one being the competency of the burner/operator. A brief, but not complete list of items/parameters that control cut quality includes:

1. Proper cutting tip selection
2. Proper cutting oxygen pressure selection
3. Proper preheat (a function of the cutting tip and the operator) this includes quantity of preheat as well as ratio of Oxygen to the fuel gas
4. Proper cutting speed
5. Proper tip to work
6. Proper selection of the fuel gas
7. A gas supply system (both fuel gas and Oxygen) that is sufficient to supply the quantity of gas required at the pressure required.

Most of these parameters may vary depending on the material and surface condition of the plate to be cut. All of these items are controlled by the operator and emphasize the importance of a well trained operator as well as a proper and well maintained machine.

References

- What-is-laser-cutting: wisegeek.com, Retrieved 27 May 2018
- How-laser-cutting-works, application-cutting: spilasers.com, Retrieved 13 July 2018
- Work-theory-laser-cutting-machine-apple-he: linkedin.com, Retrieved 30 June 2018
- Laser-Dicing-Technique-Cuts-Wafers-from-the-31907: photonics.com, Retrieved 15 March 2018
- What-is-flame-cutting: esabna.com, Retrieved 25 June 2018

Permissions

All chapters in this book are published with permission under the Creative Commons Attribution Share Alike License or equivalent. Every chapter published in this book has been scrutinized by our experts. Their significance has been extensively debated. The topics covered herein carry significant information for a comprehensive understanding. They may even be implemented as practical applications or may be referred to as a beginning point for further studies.

We would like to thank the editorial team for lending their expertise to make the book truly unique. They have played a crucial role in the development of this book. Without their invaluable contributions this book wouldn't have been possible. They have made vital efforts to compile up to date information on the varied aspects of this subject to make this book a valuable addition to the collection of many professionals and students.

This book was conceptualized with the vision of imparting up-to-date and integrated information in this field. To ensure the same, a matchless editorial board was set up. Every individual on the board went through rigorous rounds of assessment to prove their worth. After which they invested a large part of their time researching and compiling the most relevant data for our readers.

The editorial board has been involved in producing this book since its inception. They have spent rigorous hours researching and exploring the diverse topics which have resulted in the successful publishing of this book. They have passed on their knowledge of decades through this book. To expedite this challenging task, the publisher supported the team at every step. A small team of assistant editors was also appointed to further simplify the editing procedure and attain best results for the readers.

Apart from the editorial board, the designing team has also invested a significant amount of their time in understanding the subject and creating the most relevant covers. They scrutinized every image to scout for the most suitable representation of the subject and create an appropriate cover for the book.

The publishing team has been an ardent support to the editorial, designing and production team. Their endless efforts to recruit the best for this project, has resulted in the accomplishment of this book. They are a veteran in the field of academics and their pool of knowledge is as vast as their experience in printing. Their expertise and guidance has proved useful at every step. Their uncompromising quality standards have made this book an exceptional effort. Their encouragement from time to time has been an inspiration for everyone.

The publisher and the editorial board hope that this book will prove to be a valuable piece of knowledge for students, practitioners and scholars across the globe.

Index